Is the Telecommunications Act of 1996 Broken?

If So, How Can We Fix It?

AEI STUDIES IN TELECOMMUNICATIONS DEREGULATION

J. Gregory Sidak and Paul W. MacAvoy, series editors

TOWARD COMPETITION IN LOCAL TELEPHONY
William J. Baumol & J. Gregory Sidak

TOWARD COMPETITION IN CABLE TELEVISION
Leland L. Johnson

REGULATING BROADCAST PROGRAMMING
Thomas G. Krattenmaker & Lucas A. Powe, Jr.

THE FAILURE OF ANTITRUST AND REGULATION TO ESTABLISH COMPETITION IN LONG-DISTANCE TELEPHONE SERVICES
Paul W. MacAvoy

DESIGNING INCENTIVE REGULATION FOR THE TELECOMMUNICATIONS INDUSTRY
David E. M. Sappington & Dennis L. Weisman

UNIVERSAL SERVICE: COMPETITION, INTERCONNECTION, AND MONOPOLY IN THE MAKING OF THE AMERICAN TELEPHONE SYSTEM
Milton L. Mueller, Jr.

TELECOMMUNICATIONS COMPETITION: THE LAST TEN MILES
Ingo Vogelsang & Bridger M. Mitchell

VERTICAL INTEGRATION IN CABLE TELEVISION
David Waterman & Andrew A. Weiss

PUBLIC POLICY TOWARD CABLE TELEVISION: THE ECONOMICS OF RATE CONTROLS
Thomas W. Hazlett & Matthew L. Spitzer

INTERNATIONAL TRADE IN TELECOMMUNICATIONS
Ronald A. Cass & John Haring

IS FREE TV FOR FEDERAL CANDIDATES CONSTITUTIONAL?
Lillian R. BeVier

TAXATION BY TELECOMMUNICATIONS REGULATION: THE ECONOMICS OF THE E-RATE
Jerry Hausman

IS THE TELECOMMUNICATIONS ACT OF 1996 BROKEN? IF SO, HOW CAN WE FIX IT?
J. Gregory Sidak, editor

Is the Telecommunications Act of 1996 Broken?

If So, How Can We Fix It?

Edited by J. Gregory Sidak

The AEI Press
Publisher for the American Enterprise Institute
WASHINGTON, D.C.
1999

. To order call toll free 1-800-462-6420 o
1-717-794-3800. For all other inquiries please contact the AEI Press, 115
Seventeenth Street, N.W., Washington, D.C. 20036 or call 1-800-862-580

LIBRARY OF CONGRESS CATALOGING-IN-PUBLICATION DATA

Is the Telecommunications Act of 1996 broken?: If so, how can we fix it? / edited by J. Gregory Sidak.

p. cm.

1. United States. Telecommunications Act of 1996.
2. Telecommunication—Law and legislation—United States.
I. Sidak, J. Gregory.
KF2762. 115.S57 1999
343.7309'94—dc21

99-29682
CIP

ISBN 978-0-8447-4094-2

1 3 5 7 9 10 8 6 4 2

The AEI Press
Publisher for the American Enterprise Institute
1150 17th Street, N.W.
Washington, D.C. 20036

Contents

FOREWORD vii
Christopher C. DeMuth and J. Gregory Sidak

CONTRIBUTORS ix

1 THE LIGHT AT THE END OF THE TUNNEL *V.* THE FOG: DEREGULATION *V.* THE LEGAL CULTURE 1
Reed E. Hundt

2 PANEL DISCUSSION 11
Robert E. Hall, Paul W. MacAvoy, and Robert D. Willig

3 SMOKE DETECTION: CLEARING THE AIR ON LOCAL COMPETITION 25
Raymond W. Smith

4 ABOLISH THE FCC AND LET COMMON LAW RULE THE TELECOSM 33
Peter W. Huber

5 THE TELECOMMUNICATIONS ACT AND ITS INFANCY: A REVIEW OF THE FIRST EIGHTEEN MONTHS 50
Thomas J. Bliley, Jr.

6 THE RACE FOR LOCAL COMPETITION: A LONG-DISTANCE RUN, NOT A SPRINT 55
Joel I. Klein

7 DAMN THE TORPEDOES—FULL COMPETITION AHEAD! 69
William P. Barr

8 PUTTING "PEOPLE" IN THE PUBLIC INTEREST 77
F. Duane Ackerman

9 FOXES, HEDGEHOGS, AND FEDERALISM: STATES IMPLEMENT THE TELECOMMUNICATIONS ACT 86
Bob Rowe

10 OUT OF THE COURTS AND INTO THE MARKET: WOULDN'T IT BE GREAT? 100
John D. Zeglis

11 CONSUMERS WANTED COMPETITION, BUT SO FAR IT'S NO CONTEST 112
Richard D. McCormick

Foreword

The Telecommunications Act of 1996 was the first major overhaul of communications policy in the United States in over sixty years. By mid-1997, however, it had become clear that considerable disagreement existed over whether the new legislation was accomplishing its purposes. On August 14, 1997, Reed E. Hundt, then chairman of the Federal Communications Commission, delivered a speech at AEI in which he proposed several detailed amendments to the Telecommunications Act of 1996. Three distinguished economists—Robert E. Hall of Stanford University, Paul W. MacAvoy of Yale University, and Robert D. Willig of Princeton University—commented on Chairman Hundt's speech and gave their own opinion on whether the local and long-distance provisions of the 1996 act were hopelessly gridlocked.

The strong response to Chairman Hundt's speech and the discussion by Professors Hall, MacAvoy, and Willig persuaded us to continue the discussion through a series of speeches addressing the questions, Is the Telecommunications Act of 1996 broken? If so, how can we fix it? The purpose of the series was succinctly to present the diverse views of state and federal regulators, telecommunications executives, public policy scholars, and legislators.

This book compiles those speeches. Each speech appears in chronological order, with its date of presentation noted. Over the course of the series, various business transactions occurred in the telecommunications industry, and key court decisions interpreting the Telecommunications Act of 1996 were issued. To preserve the contemporaneous feel of the speeches, and to avoid the Sisyphean task of trying to keep pace with an intensely dynamic industry, we have not edited the speeches to reference subsequent developments.

Publication of these speeches should serve a valuable intellectual purpose. They memorialize what influential leaders in industry, government, and academia thought about an important piece of deregulatory legislation early in the process of its implementation. Following Chairman Hundt's speech, all speakers addressed the same questions. Their responses, even when polemical, provide a valuable historical record for future policy research. The speeches also have substantial pedagogical value. We hope that professors who teach courses on business regulation and strategy will assign these speeches to their students and ask them to make sense of the divergent viewpoints expressed.

CHRISTOPHER C. DEMUTH
J. GREGORY SIDAK

Contributors

F. DUANE ACKERMAN is chairman, president, and chief executive officer of BellSouth Corporation.

WILLIAM P. BARR, former attorney general of the United States, serves as executive vice president for government and regulatory advocacy and as general counsel of GTE.

THOMAS J. BLILEY, JR., represents the Seventh Congressional District of the Commonwealth of Virginia in the U.S. House of Representatives. He is chairman of the House Committee on Commerce. Mr. Bliley led the passage of the 1996 Telecommunications Act.

ROBERT E. HALL is a senior fellow at the Hoover Institution and professor of economics at Stanford University.

PETER W. HUBER is a senior fellow at the Manhattan Institute for Policy Research.

REED E. HUNDT served as chairman of the Federal Communications Commission from 1995 to 1997.

JOEL I. KLEIN is assistant attorney general of the Antitrust Division of the U.S. Department of Justice.

PAUL W. MACAVOY is the Williams Brothers Professor of Management Studies at the Yale School of Management. He is the principal academic adviser to AEI's Communications Deregulation Project.

RICHARD D. MCCORMICK is chairman of the board of U S WEST, Inc.

BOB ROWE is the first vice president of the National Association of Regulatory Utility Commissioners and a commissioner at the Montana Public Service Commission.

J. GREGORY SIDAK is the F. K. Weyerhaeuser Fellow in Law and Economics at the American Enterprise Institute for Public Policy Research and a senior lecturer at the Yale School of Management. He directs AEI's Communications Deregulation Project.

RAYMOND W. SMITH is chairman of Rothschild North America, Inc. He retired as chairman of Bell Atlantic Corporation on December 31, 1998, and was chief executive officer of the corporation through June 1, 1998.

ROBERT D. WILLIG is professor of economics and public affairs at Princeton University. He is also the principal external adviser to the Inter-American Development Bank.

JOHN D. ZEGLIS is the president of AT&T.

Is the Telecommunications Act of 1996 Broken?

If So, How Can We Fix It?

1

The Light at the End of the Tunnel *v.* the Fog: Deregulation *v.* the Legal Culture

Reed E. Hundt

A year and a half after the president signed a law that replaced 100 years of monopoly in communications with a commitment to competition, we should ask, "Is it working? Will Congress and the president see their intentions come true? Will Americans get competition and deregulation in local telephone markets, the largest remaining monopoly in our economy?"

The answer is: We have scarcely any competition in the local markets. The pace of investment in new entry is excessively slow, and the delays, missteps, and complexities of our legal culture jeopardize the success of our country's national deregulation effort. Here, I describe some of the problems that legal culture poses for rapid deregulation and some of the solutions that are possible.

Jerry Mashaw of Yale Law School wrote in his brilliant book analyzing the failures of the 1966 Motor Vehicle Safety Act: "Legal culture is expressed through the operation of judicial review, the separation of powers, federalism, and assorted checks and balances." And that legal culture, he wrote, determines the ultimate success or failure of any law that depends, as the Telecommunications Act must depend, on agency implementation.

As Mashaw explained, the legal culture hornswoggled the congressional intent behind the unanimously passed Motor Vehicle Safety Act. While the courts debated for decades every nuance of every phrase of the law, and every jot and every tittle of the expert agency's implementation, almost every year since that law was passed more Americans have been killed on the highways than in the entire Vietnam War.

When it comes to legislation, between the thought and the deed falls the shadow. The shadow is cast by the million lawyers of America. I praise, respect, and try to convince other countries to adopt the American system of rule of law, but respecting the rule of law is different from admiring the delays of the legal process.

The Telecommunications Act described a paradise of open markets, free competition, deregulation, and growth in the information sector that is already one-sixth of our economy. I view Congress and the Federal Communications Commission together in that paradise like Adam and Eve, the FCC being the agency created by Congress like Eve—from Adam's rib. But now the FCC and Congress, like the first family of Milton's *Paradise Lost,* are being exiled hand-in-hand with slow wandering steps from Eden into the harsh desert of the legal culture, with its thousand devices of tortuous delay and tortured questioning of every phrase, word, and punctuation mark of the act.

Professor Mashaw wrote that an agency must "understand how hostile the legal culture can be toward certain forms of regulation." I can testify how hostile the legal culture can be toward deregulation. I admit that Congress asked the FCC to undo the largely state-regulated local monopolies by means of forceful rulemaking, not by passivity or mere jawboning.

But hostility is a fair way to characterize the reaction from the telephone industry, from many states, and even from alleged new entrants. The long-distance companies were hostile when the FCC dared order them no longer to file tariffs with the commission. But if they were not regulated, they might be more exposed to consumer lawsuits, so they got a court to enjoin the FCC from deregulating.

I understand that for incumbent monopolists and other companies protected by regulation, hostility might be a rational response to deregulation. The problem is that the legal culture, just as the professor taught, validates that hostility, primarily by encouraging unceasing argument and ineffective, delay-ridden decisionmaking. Debates of interminable Jesuitical casuistry rage over the congressional intent behind the word *cost,* but my favorite is the debate over the word *and,* as in the phrase *business and residential.*

I admire the lawyerly zeal that sees complexity behind those terms. Of course, I do so because I am a lawyer; worse than that, I am a son of a lawyer. So I appreciate lawyerly cunning that can fog the plainest congressional intent. But how can we find our way to free markets through the lawyerly fog? Even now, the local operating companies and the long-distance companies and many other parties, often including the states, appeal virtually every paragraph of every one of the FCC's decisions to various courts of appeals. And the courts do not consolidate the appeals; they scatter them like grain in the wind across the country.

As if there were not enough litigation, Southwestern Bell has hired a brilliant lawyer named Laurence Tribe to argue in federal court that the Telecommunications Act that it lobbied about for years is unconstitutional. I have two observations.

First, has SBC just discovered that the Telecommunications Act is unconstitutional? The firm lobbied about the act for a decade, and a year and a half after it was signed, SBC just found out that it violates the Constitution?

Second, the Bells asked Congress for a date by which they could enter long-distance markets. They wanted three years, which would have been February 1999. And they lost that lobbying effort. Now they are irate to the point of starting tribal warfare because halfway through their own proposed waiting period, they have not yet made a record of opening local markets in any state that passes the test of the law according to the Department of Justice. The fog, the fog.

Meanwhile, after getting the Eighth Circuit to enjoin the FCC's clear national rule defining *cost* so that fair and final interconnection agreements could be quickly reached, GTE's taste for litigation is unslaked. So far, GTE has filed thirty federal court actions challenging the pricing decisions of twenty-three state commissions.

Some say that the Supreme Court will eventually resolve all disputes and that the path to competition will be clear. And that might be true, but the excruciating delays of the legal process recall the lawsuit known as *Jarndyce* v. *Jarndyce* in Charles Dickens's *Bleak House.* That case could just as easily describe an interconnection suit. Dickens wrote:

> This scarecrow of a lawsuit has in course of time become so complicated that no man alive knows what it means. The parties to it understand it least, but it has been observed that no two lawyers can talk about it for five minutes without coming to a total disagreement as to all the premises. The little plaintiff or defendant who was promised a new rocking horse when the case should be settled has grown up, possessed himself of a real horse, and trotted away into the real world.

Well, I myself perhaps will find that horse and ride away, the Senate so permitting when it returns in September, and so will many, while the interconnection litigation drones on. But there has to be a better fate for our country's competition policy than *Jarndyce* v. *Jarndyce.*

Dickens also wrote, "Facts alone are wanted in life, when nothing else, and rule out everything else," so I would like to plant a handful of facts with you.

Fact number one: After years of effort by states such as Ohio, Wisconsin, Illinois, Oregon, and others before Congress passed the Telecommunications Act, and a year and a half after the act was passed, consumers still have scarcely any choice for telephone service in local markets. New entrants have, at the most, about 1 to 1.5 percent market share, according to all accounts. So far, scarcely any local competition has been delivered to residential or business customers.

Fact number two: If our local telephone companies thought that markets were truly open, it stands to reason that at least one of them somewhere would be going into its neighbor's market. For all intents and purposes, virtually none is doing so.

And if local markets were truly open to competition, then all Bell companies in those markets would long ago have filed applications for entry into long-distance markets. Instead, only one Bell in one state—Ameritech in Michigan—has tried to make a serious showing that the state is truly and fully open to competition, and in that case neither Michigan nor the Department of Justice agreed.

By the absence of serious Bell operating company entry into interLATA services and by the absence of competition among local telephone companies, the local telephone industry in effect admits that not one state market is truly open to competition at this time.

Fact number three: The congressional intent to open local markets on a national level has never, not for a day, been completely put into effect, and that has mattered a great deal. Since the day the law was signed, incumbents have been arguing for prices so high for interconnection that new entrants would, in effect, be paying off the incumbents' past investments. If new entrants do that, they will not have enough money to pay for their own investments, and the result will be no competition.

A new entrant could never be assured in any definitive, final manner that interconnection pricing and the pricing of network sharing—of unbundled network elements—would be fair in any state or any region of the country. In other words, a new entrant could never count on certainty that across a region and across this country it would get forward-looking pricing—pricing set by the formula based on total element long-run incremental cost. Indeed, at last count only eleven states of fifty and the District of Columbia have even issued permanent interconnection and element pricing rules.

Does it matter that the Eighth Circuit enjoined the FCC's fair national pricing methodology from going into effect? GTE told the Eighth Circuit last fall that if the FCC's pricing rules went into effect, GTE would suffer "substantial and rapid losses of market share." In other words, GTE said that competition would exist right away. GTE got an injunction from the Eighth Circuit, and the prospect for rapid competition went away. Result: The local telephone companies have lost hardly any market share since the law was passed. So the judicial intervention in the interconnection case hurt competition badly.

Fact number four: The existing competitive efforts in local markets are tiny fish that will not survive in the presence of the incumbent, formerly government-protected, monopolistic, whale-sized local telephone operating companies, not unless state and national governments write and enforce procompetitive rules. All evidence suggests that as of now enforcement is not effective because the means of enforcing binding competition rules at the state or federal level do not exist, and without that, competition will be undercut.

Those four key facts tell the FCC that it has a major challenge to introduce competition in local telephone markets and that the challenge is not yet being met. Now, plainly, investment in local competition is the goal. Investment in building facilities-based competition is the goal, and the commission needs to promote investment by promoting wise and rapid decisions about how to open local telephone markets. Justice delayed is investment denied.

Now I get to my most comforting fact, number five: The Telecommunications Act is not the Ten Commandments. The commandments were brought down directly from the Supreme Being. They were not subject to judicial review. The tribes of Israel did not have divided jurisdiction over enforcing them.

But legislation has this to say for itself: It is the work of mere humans, but mere humans can keep writing and rewriting it until courts finally submit to its direction.

So I would like to talk about how Congress could blow away the paralyzing fog the legal culture is spreading over the legislative and executive mandate to open the local markets—how Congress might truly realize its own high intent. I have four proposals for legislative action.

First, we need an immediate national definition of *cost* to be set forth by the FCC so that key term will not vary confusingly and uncertainly and unreliably from state to state and region to region. Until the Eighth Circuit interconnection decision, no court in sixty years had ever constrained the FCC from defining terms in a congressional statute, and undoubtedly Congress intends for someone to provide a single national meaning to the term *cost.* The members put the term in the statute, and congressional statutes do not mean different things in different states.

The result of the Eighth Circuit decision denying the FCC the ability to write a rule giving that definition is that as many as ninety-three federal district courts will define the term in the context of reviewing potentially hundreds of interconnection agreements. Then, a dozen courts of appeals will review those decisions, and if there is conflict among circuits—and that certainly seems likely—years from now the U.S. Supreme Court will decide whether it prefers historic cost, long-run incremental cost, total service long-run incremental cost, or any other of innumerable variations. What fog that process will spew forth!

Senators Lott, Hollings, Stevens, and Inouye and Representatives Bliley and Markey told the Eighth Circuit that the FCC had the authority to define *cost.* The words of that bipartisan group should have been treated by that court as law; they were not. Now, the law should be written to instruct that court to follow the guidance of the Senate majority leader and his bipartisan group by permitting the FCC to define the term and to have its definition respected by the courts, the states, and the parties.

Second, the FCC, right or wrong in its judgment, should be given the deference the Supreme Court authority requires for its interpretations of congressional intent and its application of policy judgments. In the 1984 case *Chevron* v. *Natural Resources Defense Council,* the Supreme Court said that where Congress has not directly addressed the precise question at issue, the appellate court should not simply impose its own construction on the statute. Instead, the court said, an appellate court has to affirm a federal agency, if the agency's answer is based on a permissible construction of the statute. We might be wrong; we might be right; but if our construction is permissible, the agency should be affirmed.

In the Eighth Circuit, by contrast, the court of appeals substituted its judgment for the FCC's when it struck down what it called the "pick and choose" rule. The commission called it the nondiscrimination rule. That is the rule that talks about whether a new entrant can obtain the same terms of an interconnection agreement obtained by some other new entrant.

The commission's view was that nondiscrimination principles required that the answer be yes. A new entrant should be able to get the same price for its agreement as that of some other entrant. That would level the playing field. The court of appeals acknowledged that the FCC's approach could reasonably be said to be intended by Congress. If that is so, then under *Chevron* the commission should have been affirmed.

But then the Eighth Circuit went on to insert its opinion for the FCC's regarding how interconnection negotiations were likely to proceed, and, apparently indulging in game theory, the court selected another approach and reversed the FCC. That decision, the commission will tell the Supreme Court, violates *Chevron.* But my point is that the decision is also a recipe for delay, doubt, and uncertainty.

Any time the FCC reasonably interprets the Telecommunications Act, the courts should rapidly affirm, not deliberately second-guess. Therefore, Congress should specifically instruct the courts that where the FCC's interpretation of the statutes is reasonable, all courts reviewing the commission's decisions should strictly follow *Chevron.* If Congress does not like those interpretations, it can change them, but the courts should follow *Chevron.*

Third, judicial review of FCC decisions interpreting and applying the Telecommunications Act should not be piecemeal or delayed. I am confident that the FCC ultimately will prevail in the Supreme Court on the Eighth Circuit's decision regarding the commission's rulemaking authority, but similar litigation over the 1992 Cable Act ended in July 1997. That took five years, and it might have gone on for another year if the cable operators had not decided to give up their chance at Supreme Court review.

We need a faster, cheaper route for getting the legal issues of competition resolved, so I propose that Congress pass a law consolidating all appeals of section 252 interconnection agreements and all reviews of FCC decisions in a single court, bar judicial stays of commission orders that last more than thirty days, and require expedited briefing and decisionmaking for all appeals and reviews of FCC decisions.

Fourth, no state or federal competition rule will be of value unless it is enforceable, and that means that it has to be promptly enforceable and coupled with appropriate sanctions.

I want to applaud the new Illinois statute that requires resolution of interconnection-related complaints within sixty days. I propose that Congress impose a similar rule on the FCC and all states. Those states that do not wish to follow such a rule should be told that their enforcement powers will then be relinquished to the FCC.

In addition, I would ask Congress to increase the sanctions available to the FCC. At a minimum, the commission should have the authority to grant treble-damage relief plus attorneys' fees in addition to equitable remedies for all violations of FCC rules. Congress, its expert agency, the FCC, the administration, and the country all want a private, investment-built information economy in which the states and the federal government have no further need to regulate investment or wholesale prices or retail prices.

Lawyers, including both judges and recovering litigators like myself, need to learn from business people the real-world consequences of delay and uncertainty. While courts spar with Congress and its expert agency, and while the National Association of Regulated Utility Commissioners sues the FCC and never, ever, I point out, files a brief that supports the commission, the fog of uncertainty spreads like the famous fog of London described by Dickens as his metaphor for *Jarndyce* v. *Jarndyce.*

And in that fog an entrepreneur misses a window for launching a new satellite venture. In that fog a would-be new entrant watches an incumbent obtain first-mover advantage in offering a new product line. In that fog a banker agrees to finance an established business while turning down money for an entrepreneur on the ground that the regulatory or deregulatory landscape is not yet clear.

The result: Jobs are lost, money is wasted, choice does not come to the market, consumers are disappointed, and productivity gains are not realized. While other economies surge forward, the United States loses its competitive edge. Those are real costs. They must always be in our minds as we lawyers try to create a fair but efficient rule of law. The competition policy of Congress should be realized by business people solving economic and technological problems. It should not be tangled up by lawyers solving or creating word games about legislative intent.

Samuel Johnson said that people disagree chiefly on means, not ends. I am quite sure that we all agree on the ends of deregulation and competition, so let us work together in mutual respect to find more effective means.

Delivered August 14, 1997.

2

Panel Discussion

Robert E. Hall, Paul W. MacAvoy, and Robert D. Willig

This discussion immediately followed Reed Hundt's assessment of the Telecommunications Act of 1996. J. Gregory Sidak moderated the discussion.

MR. SIDAK: I have asked each of our three world-renowned economists—Robert E. Hall, Paul W. MacAvoy, and Robert D. Willig—to discuss the Telecommunications Act of 1996, particularly in reference to issues raised by Reed Hundt.

MR. HALL: Chairman Hundt has introduced fact lists to the discussion, so let me start with a fact list that is much more global and grandiose that any of his. I particularly want to address whether the United States is in danger generally, as he suggested, of becoming mired down in a national *Jarndyce* v. *Jarndyce* syndrome. I think that is not so.

First of all, the United States has the highest level of output per worker in the world. By any reasonable measure, the United States has the most successful economy in the world. A very important ingredient in the success of the U.S. economy is our nation's effective rule of law. My recent research shows that more than anything else, the differences between poor countries and rich countries result from their rules; rich countries have good rules and enforce them. The rules work and cause people to focus on business activities instead of diverting value from others. In that respect the United States has the most successful economy.

Intrinsic to our successful economy is the reality that the United States has by double the largest number of lawyers per capita and the largest volume of litigation. That is how our system works, and it is good when judged against any other system. While our system has some defects, comparative analysis of 133 countries shows that our system is the best in terms of its final, bottom-line results.

Chairman Hundt mentioned cars and auto safety in a way that puzzled me. The United States has, compared with any other country in the world, half the level of mortality per passenger-mile, even in comparison with Canada. Although the Motor Vehicle Safety Act has been the subject of much litigation, it has been a very successful effort. In addition, the number of people who are killed on U.S. highways today is half what it was in the 1950s and far below the number of people killed in Vietnam.

Finally, with respect to the Telecommunications Act of 1996, the United States has, as we take for granted, the best telecommunications system of any country. As those who travel abroad know, the U.S. telecommunications system is so good that an active market of people now places long-distance calls within Europe for which the transport occurs in the United States. The call goes across the Atlantic and back again, because it is cheaper to call through the United States than it is to make a call of even a few hundred miles within Europe.

With respect to the issue of competition and the Telecommunications Act's desire to create more effective competition, one very important thing to say is that we already have a lot of competition. Competition does work in telecommunications. We have a very effective competitive long-distance market in which relatively little intervention occurs. There is not a lot of litigation with respect to the existing players in long-distance markets. We have growing, effective competition in local toll markets. One of the great tragedies of the Telecommunications Act was that it prohibited the extension of competition in local toll through the prohibition of local-toll presubscription. So, with respect to the fostering of competition, the Telecommunications Act has largely not addressed a huge negative factor.

A second point I would make is that the telephone network requires a lot of cooperation. The cooperative relations that exist today between local carriers and long-distance carriers in particular are a model of effective cooperation. Cooperation has been effective primarily because of the genius of William F. Baxter, who anticipated the problems that the Telecommunications Act has brought. He said, "If you want competition, let's have competition horizontally but prevent vertical integration, which is an enemy of cooperation." So we have a highly competitive long-distance industry, which under the 1982 system is the customer of—and not a competitor to—the local telephone companies. The structural separation principle that was embodied in telephone policy from the beginning of 1984 until the Telecommunications Act has been a huge success where seriously applied—in long-distance markets. Again, the improvement in performance of the U.S. long-distance industry, which I have documented in other work, is amazing in comparison with the performance of almost any other industry.

So the issue of how policymakers manage cooperation as they bring more competition into telecommunications is really the central issue. That is the issue that the authors of the Telecommunications Act did not take seriously. The Telecommunications Act relies not on the principle of structural separation to induce cooperation, but rather on the principle of forcing cooperation among competitors. That is the reason the act has gotten enmeshed in the kind of value-dissipating activities that Chairman Hundt described—ones that I agree are a tragedy.

We hear more and more discussion of how we could extend the principle of structural separation. The most promising way in my view would be to impose separation at a very different level—namely, a separation of the local loop from the rest of the system, in which we would retain a regulated local loop and move immediately to competitive local transport and retain, obviously, competitive long-distance markets. Analysts have well established that the Telecommunications Act's approach does not work. We need to come up with systems that recognize that policymakers can avoid dissipative litigation by intelligently applying the principle of structural separation.

MR. MACAVOY: Robert Hall says that the Telecommunications Act does not work. I believe that everyone interested in this topic must have some basis for taking that position.

The Federal Communications Commission is having its Vietnam. The commission is at the same point as was the Department of Defense in 1970 when it tried very hard to redefine the war so that the body count would be the single measure of victory. Chairman Hundt wants to create binding competition rules as the equivalent to the body count. But in the same way that there was no such thing as a dead soldier, there is no such thing as a binding competition rule. Competition is not a set of binding rules; it is interactive behavior among firms in open markets, having relationship not to the size and age of the firm but to relative cost and the ability of entrepreneurs to change the product in ways that favor consumers or that consumers favor.

With that format, the position of the FCC cannot be changed in either the short or the long run. To begin to explain it all, let us go back to 1994 and to the records of one of those interminable commission proceedings, with tens of thousands of pages of submissions. You will find that one Bell operating company presented a document that showed the revenue sources for classes of services to system subscribers in one state in the Midwest. I would think that a potential competitive entrant into the local exchange market would be interested in that document. The document showed thirty lines of business and revenues by lines of business for 1994, with $200 million for a monthly average, for service in three regions: a major metropolitan region, the suburbs of that metropolitan region, and the generally rural or agricultural rest of the state. Of the types of service, however, Centrex, local access, switched interstate access, and residential and business usage provided revenue flows of about $154.4 million. The document provided an estimate of incremental costs with each source of revenue.

That information is interesting for answering a potential question central to competitive entry conditions: After paying for the cost of operation of that facilities-based network, how much would be left over for what one might call a "contribution" for that class of service? What is the operating income margin on those four most important services?

Before answering that, I would think that one would find it difficult to make a case to shareholders for entry if the operating income margins were below 30 to 40 percent. The infrastructure to provide for the set of services—advertising, organization, technology, and research—would require a contribution in the range of 30 to 40 percent at the least.

The Centrex margin for that operating company in 1994 was 41 percent. The local access margin was 23 percent. The residential and business per minute usage margin in the high-density metropolitan region was 49 percent. That margin should be broken out, however, because it was about 65 percent for business, which provided one-third, and about 45 percent for residence and business usage, which was the rest of the $62.4 million. The switched interstate access margin was 75 percent of sales revenues.

If I were a potential competitive entrant into that state, I would go to the board of directors with a strategic plan and make the following argument: "We cannot do it in house-to-house local access. We may have to provide some targeted local access to provide business usage, but we should target our entry on the basis of business usage in the metropolitan region and switched-interstate access."

How does all that relate to the goals of the FCC? Suppose that a second firm is sufficient to provide competition in local access and exchange in such a market. Then, I would find that, having a favorable decision of the board of directors, the potential entrant would produce a business plan that would provide for service to zone A business, where the profit margin is 65 percent; for custom calling, central office services for those businesses, where the profit margins are 90 percent; and for switched access for interstate, where the profit margins are 75 percent. The entrant would never provide residential service in that state.

The FCC would not find that to be competitive entry in local exchange, as required for step one of deregulation under the 1996 Telecommunications Act. Such entry is not what any of us who practice the analysis of economics calls "competitive." It is a result for a submarket or a set of submarkets. Indeed, since 1994 in that particular set of submarkets that could be profitable, entry has occurred, and the incumbent has lost a third of its share. Prices have gone down for those particular services, probably in current dollars by 15 to 20 percent. The rest of the state is just the way it was in 1994.

The basic argument, then, has to be that a case does not exist for structural establishment of a second, third, fourth, or fifth source of supply in local exchange. The "competition" from deregulation that has developed in a limited way in the airline, trucking, and railroad industries will not occur. Competition will never happen, regardless of the presence of the Federal Communications Commission.

MR. WILLIG: What particularly caught my attention in Chairman Hundt's talk, besides the phrase *local competition,* which is certainly of great interest to all of us, was the phrase *legal culture* and also the word *process,* because those are not ordinarily the domain of microeconomists, although we do like to be imperialistic from time to time. All of us on this panel sometimes allow our minds to drift into somewhat more difficult and strange waters than mere economics. Thus, Chairman Hundt's remarks hit some of my own passions on the subject, so I am delighted to give the economist's version of some of his points. To me, it seems, as it does to Chairman Hundt, that the culture of telecommunications policy today is highly troubled. And I chose the words *culture of telecommunications policy* carefully. While you gentlemen were at Yale, the word *process* was all over my sociology class notebook, along with *culture.*

What undergirds culture and process, I was told in that course, is language. It strikes me today that our vocabulary describing local competition, as well as legal culture, has turned positively dysfunctional. Traditional words are in danger of losing their meaning, even in the mouths of people on this very panel, so that logical propositions that are stated in that vocabulary in fact sound and have become upside-down. That strikes me as a cultural process-laden phenomenon, and Chairman Hundt has indeed introduced that perspective on the subject.

Chairman Hundt actually used the word *deregulatory* to describe the 1996 act. We all have probably used that word at one time or another. None of us is immune. The voters, the press, our political representatives, even professors—all seem to be waiting for deregulation to occur.

We keep thinking, "Well, of course, the FCC must act first, with congressional and academic guidance and perhaps some from the Department of Justice. Yes, those are regulatory bodies intrinsically, but perhaps soon, if the regulators act well and speedily, we shall turn the corner, and deregulation will be here, local competition will be here, the entrepreneurs will come out of the fog, and all will be well in the industry. As Robert Hall points out, the rest of the economy is doing just fine.

That all seems to me a terrible abuse of vocabulary. Instead of blaming our disappointment and frustration on the lawyers or on the government—certainly not on the economists—we need to look inside and tell ourselves the truth: The vocabulary—the culture of policy thinking today—is really very much at fault as well. That is a matter of very simple economics.

The central fact is that the local exchange is still a monopoly. In fact, it still fits all the classical economic characteristics of a natural monopoly.

Just a few months ago, some people still believed that any day the competitive access providers—the cable TV companies, the electric utilities, the wireless companies—would be providing us with facilities-based competition. We were really all relying on those dreams. I do not hear that sort of talk anymore. We have all come to realize that facilities-based competition is not just around the corner. Maybe in two years, maybe in five years, but we have no particular reason to have high hopes for an immediate solution from that corner. Competitive access would be nice, but it is not happening.

The second fact is the enormously powerful economic incentive to lever monopoly power over local exchanges into any connected market for which local services are a bottleneck. Those incentives certainly apply to long-distance service as well. Regulation created those incentives, yet we truly need regulation over the monopoly bottleneck.

And the third fact is, of course, that companies have enormous ability to degrade quality at the interface, to make life hard for their rivals by abusing local bottlenecks. Given the powerful economic incentives, however, a fear of government power—in the form of regulation or antitrust litigation—is the only thing that might stop that very profitable bottleneck abuse. One might hope for reputation, for public opinion perhaps, for consumers' favorable views of companies somehow to rein in those incentives. But we all understand that that is just not a powerful enough set of forces, certainly under the kind of vertically disintegrated marketplace toward which we are lurching, because it is impossible for consumers to know whose reputation the failures of quality and reliability ought to muddy.

Despite those three problems, we all yearn for some simple solution. That is an interesting and subtle strain in our culture. We want to believe that voluntary negotiations will solve our problems.

If we had no regulation and we were to turn to voluntary negotiations, the fall-back position for holders of local bottlenecks would be full monopoly profit. Voluntary negotiations without regulation might lead to a cooperative agreement, but only one that would confer those same fall-back levels of monopoly profit on the holders of a local bottleneck. That is not a solution that anyone who rings the bell of voluntary negotiations truly wants. Economics says that voluntary negotiations can be wonderful within a regulated structure whose first purpose is to pass along to consumers the fruits of what otherwise would be monopoly profit.

The bottom line is that it is high time for us collectively, as a culture, as a policy community, to bite the bullet and say, "Yes, we do need a regulated solution." That is what the challenges of this marketplace force on us, despite the fact that we are aware of the costs of regulation.

We need a good regulatory solution. We need one that is able to allow market forces to operate competitively, without the distortion of bottleneck abuse. But we are not going to get there until the members of Congress, the FCC, the press, the voters, and economists recognize that what we need is an intelligent regulatory system. We must stop the fantasies of a market solution or a voluntary negotiation without the hand of regulation guiding it.

With all of that said, I appreciate the remarks of Chairman Hundt. I am not sure that I really wish to confer all that authority, as a matter of public policy, on the FCC. That notion is a little scary. But the basic thrust of his remarks is very well taken, and I consider Congress and the courts part of the problem at this point.

I would add one topic to the list of items that Chairman Hundt mentioned for congressional attention. We need, in this imagined new legislative supplement to the 1996 act, some clarity on the role of the Justice Department and on how that role is to be translated into an assessment of the state of the market. Some members of Congress are trying to hold back the Justice Department. I trust the Justice Department to put some teeth into the instruction to ensure that the regional Bell operating companies open the market to competition. Competition is very much an appropriate subject for the expertise of the Justice Department. Of course, the FCC can lend its expertise as well. It would be extremely dangerous for congressional pressure to stand in the way of that proper consultative role. One has to keep one's eye on competition.

MR. SIDAK: Let me describe three scenarios and ask the panelists to assign probabilities for each. For purposes of the question, assume that entry into the local market and entry into the interLATA market are gridlocked. Ameritech has been trying since at least the early 1990s to negotiate terms to get into interLATA markets. But that has not happened yet, and it is not going to happen next week, according to Chairman Hundt. So I think we can say that the in-region interLATA entry process is a failure. Meanwhile, Paul MacAvoy's remarks about entry into the local market suggest that nothing is happening there very quickly in residential and business access service, although higher levels of entry may occur in certain high-margin services. In addition, we observe the FCC's tendency to put off to a future day any resolution of issues relating to cost recovery in pricing dockets, access reform dockets, or universal service dockets in such a way that we never get to the takings issue.

The first scenario is "mutually assured destruction." All the local exchange carriers sue the states and the FCC on takings grounds. The interexchange carriers retaliate by suing all the local exchange carriers on antitrust grounds for perpetuating their local monopoly.

The second scenario is "detente": The regional Bell operating companies, GTE, and the other local exchange carriers make various vertical mergers with AT&T, MCI, Sprint, and WorldCom. The pricing disputes over access to the local network, as well as the disputes over entry into the interLATA market, become academic and internalized. They are resolved within the companies, perhaps with the helping hand of the FCC and the Justice Department, maybe along the lines of what was negotiated with Bell Atlantic/NYNEX.

The third scenario is "velvet divorce." Here, the local exchange carriers spin off their operating companies completely and create a separate entity that will hold all unregulated or non–local exchange carrier activities. Those non–local exchange carriers are then able to buy resale and unbundled elements on the same pricing terms as AT&T, MCI, Sprint, and others. Then, those pieces of the former local exchange carriers immediately enter the interLATA market. The local exchange operations remain as some highly regulated activity that cannot get engage in other lines of business.

What are the probabilities of each scenario? If the panelists can see a fourth or fifth scenario, please elaborate. But remember that your probabilities have to add up to 100 percent.

Also give me a specific date at which a regional Bell operating company, in either its current or future incarnation, will first commence provision of in-region interLATA service.

MR. HALL: I have a partial answer and a partial amendment to your scenarios, because they all sound like fun for me. But I have a fourth possibility—the one that keeps me up at night. That is that somehow, again because of the cultural distortions in the policy process, we tire collectively of struggling and essentially give in to what the press seems to be touting as the easy route. That route is to pretend that conditions like those of Bell Atlantic/NYNEX are sufficient to ensure that we do have a misuse of the phrase *competitive marketplace* such that the regional Bell operating companies are allowed into interLATA territory. Everyone smiles and says, "Ah, at last we've done a good job." Then, after a few years of apparent peace in the industry, all hell breaks loose. Maybe we are back to the first scenario, with antitrust cases and abuses flying about, with consumers being the clear losers. I fear that that is the most likely scenario.

MR. SIDAK: So that is 51 percent?

MR. HALL: I do not know. Say, 43 percent.

MR. MACAVOY: Could I produce a set of probabilities that would associate me with not knowing what the answer is? From the point of view of a person interested in the regulated industries since I wrote my dissertation in the 1950s, I believe that looking elsewhere works. For one thing, mutually assured destruction did indeed work in the railroad industry. The Interstate Commerce Commission destroyed the industry to such an extent that the demise of the commission cost the commission nothing: it had nothing left to regulate. And when the deregulation process was complete, out of the ashes appeared a smaller, highly efficient, somewhat profitable industry with rates at half the level they were before deregulation. I do not doubt that all the players have a great deal of self-interest in maintaining court proceedings in perpetuity. The only reason I would give the first scenario less than .99 probability is that the industry is still such a wonderful source of technology and invention that it might outlast an attempt to destroy it.

The second scenario is more interesting. I give it a higher probability because it goes to the issue of what is really at stake among decisionmakers. The decisionmakers that make a difference are the twelve largest sources of supply in interLATA long-distance and local exchange markets. What do they want? Where are the extraordinary profits to be made from changing the institutional structure before us—a process that seems to have come to a stalemate?

I believe that the excess earnings in interLATA long-distance service, generated by the tacitly collusive operations of AT&T, MCI, Sprint, and WorldCom, are at stake. Lerner index levels exceed 60 percent on all high-volume services. That now-unrequited opportunity is the reason we witness the flagellation of the Bell operating companies as they desperately seek to comply with impossible requirements for an FCC license to enter long-distance markets. Those companies exhibit a willingness to do almost anything for some significant percentage of the annual excess earnings. Your second scenario is a way for the regional

Bell operating companies to do that, and SBC almost worked that out. Only the incompetence of AT&T's management prevented SBC's entry. There may be other ways. MCI is well along to becoming an integrated local and long-distance business carrier able to forestall entry into its sequestered parts of the long-distance market. Therefore, various vertical mergers that internalize a lack of cooperation may very well be worked out.

The third scenario leaves me cold. Somebody has to own local operating companies. Regulatory and natural monopoly barriers are not going to go away just because the local exchange carriers spin them off. So I do not see where that scenario goes at all.

MR. SIDAK: What if a local exchange is owned by government, just as roads are owned by government?

MR. MACAVOY: Then, that begins to look more like what is happening in the electric power industry, where government is going to end up owning the backbone network after the spinoff of delivery at retail. Nobody is going to want to own networks that will pay rates based on total element long-run incremental cost to use power grids, and they will run at losses in the tradition of the Tennessee Valley Authority. And that could happen here, too.

I do not know whether that makes the third scenario look worse than than the first one, but I think that both of them are business as usual. Operating companies going into long-distance markets have too much rent to earn for the second scenario not to happen.

MR. SIDAK: So you would say that the second scenario is the most likely of the three scenarios?

MR. MACAVOY: Money counts. Excess profits are, even under regulation, there to be dissipated in long-distance markets by competitive entry.

MR. WILLIG: First of all, I have to assign a high probability—something like 90 percent—to the first scenario including what I would see as the Willig alternative, which is that the Bell entry into long-distance markets occurs at some point during that process. Certainly, many of the other things listed in the first scenario would occur under the Willig version as well. I might add that the situation would be ripe for a consumer class-action suit, as well as for litigation among the sellers.

It seems to me that the detente scenario has a low probability. After all, it represents nothing more than the re-creation of the old AT&T, as far as any given telephone customer is concerned. It means that each customer would have only one choice for vertically integrated service. If any of the things that Robert Hall and I have written come true, which I think they would, the choices for separate purchase of long-distance service in that environment would be very limited relative to what they are today. The Department of Justice would have to be run by someone whose thinking is very different from Joel Klein's for the second scenario to go anywhere. I would give it a probability of 1 percent.

The third scenario, of course, is the favored one. I am tempted to give it a high probability simply because I think that it is a good one. The problem of who owns and operates a local loop is solvable, and it would not have to be the government. It is, as Paul MacAvoy says, just a matter of money. There is a regulated rate for the use of a local loop that would induce an owner to take responsibility for it, even though that owner would not have the right to be involved in any other aspects of the network. I continue to think that such an alternative is very workable. Among those scenarios, that is by far the best.

But do not confuse *is* and *ought.* I cannot give the third scenario a high probability just because it is the good scenario. And following the Sidak rule, I give it a probability of 9 percent, despite my enthusiasm. I think in fact that we are headed for the first scenario or some variant of it. That is too bad, for all the reasons that Chairman Hundt said. It is not only a costly process—to consumers and to parties—but enormously inefficient. But I do not see any political way to avoid it.

August 14, 1997.

3

Smoke Detection: Clearing the Air on Local Competition

Raymond W. Smith

I am pleased to address the promise—and the unfulfilled potential—of the Telecommunications Act of 1996. Bell Atlantic's commitment to implementing the letter and spirit of the Telecommunications Act could not be clearer. The firm has opened its local market by negotiating almost 300 interconnection agreements with scores of competitors. And agreements require two signatures—that of the local exchange carrier and that of the competitor, including of AT&T, MCI, and Sprint.

Before merging with NYNEX, Bell Atlantic struck a landmark agreement with the Federal Communications Commission that addresses every one of the concerns expressed by the interexchange carriers and federal regulators. Bell Atlantic voluntarily committed itself to stringent performance monitoring, uniform interfaces, flexible pricing, and forward-looking costs. No other local operating company has done so. To live up to those agreements, Bell Atlantic is investing hundreds of millions of dollars in the systems and processes required to serve local competitors. The company processes between 1,000 and 2,000 orders a day for resale lines in New York City alone and has the capacity to process five times that many if necessary. Bell Atlantic has also negotiated deeply discounted rates for resale and unbundled elements so that competitors can offer local service if they choose to do so.

The Telecommunications Act is working fine. In the year and a half since it passed, Bell Atlantic has made dramatic progress toward the act's vision of open competition in local markets. Yet, every week new, misleading ads appear in *Roll Call* and the *Washington Post* with competitors' claims that Bell Atlantic is preventing them from offering local service. The truth is that those who say that Bell Atlantic is not implementing the act are uninformed or purposely inaccurate. Their claims are meant to create a climate of dissension that impedes progress. More than that, those misstatements serve as a smoke screen to divert attention from what is really happening in the marketplace today to develop the competitive marketplace that the act envisioned. Some plain talk and plain facts can clear away that smoke.

One plain fact is that the FCC, with its vision impaired by the diversionary rhetoric of the big interexchange companies, has been looking for local competition in all the wrong places. Another plain fact is that, while some companies drag their feet and wait for regulators to subsidize prices and mandate market-share loss, local competition has bloomed—thanks to good, old-fashioned market incentives, new technologies, and the long-term business interests of savvy, entrepreneurial new entrants. A third plain fact is that Bell Atlantic has been investing millions of dollars and taking all the necessary actions to open its networks and stimulate the facilities-based and resale competition that creates investment, innovation, and customer choice. The company does so not out of altruism but because it has the business incentive to do so—more than any local operating company in the country.

I begin by laying out the economic realities—at least as they exist for Bell Atlantic. The company's number-one objective is to provide a full range of services—local, long-distance, and data—to the best, most competitive market in the country. In fact, I brought Bell Atlantic and NYNEX together for the very purpose of mining the growth opportunities in the merged entity's rich regional marketplace. The only way to do that is by being a full-service, one-stop provider. That is my commitment to Bell Atlantic customers, employees, and shareowners, and that is the reason that I feel a personal sense of urgency to get the job done and get it done right.

The clock is ticking. Already, competitors have bought more than 30,000 unbundled loops and more than 130,000 resale lines, much of that concentrated in New York City. Competitive local carriers have already taken a big chunk of Bell Atlantic's lucrative business market, with more inroads being made every day. Bell Atlantic exchanged 5 billion minutes of use with competitors in 1997—the equivalent of all the calls made in Vermont in a year. Intending to compete vigorously for those customers, Bel Atlantic needs to be able to provide long-distance service to do so.

I was astonished that Reed Hundt claimed that Bell Atlantic has no real interest in entering the long-distance market, especially when his comments occurred within days of the WorldCom bid for MCI. If that deal goes through, WorldCom will have all the assets necessary to offer local, long-distance, and data transport services—coast to coast, end to end. WorldCom has made no secret that its prime target is the Bell Atlantic region. Without long-distance relief, Bell Atlantic will be competing, with one hand tied behind its back, against a well-financed, full-service provider for the most lucrative accounts.

Perhaps Bell companies with more wide-open spaces than wide-open markets can afford to ignore a challenge like that, but Bell Atlantic cannot do so. I have a $20 billion reason for wanting to be in the long-distance business. Bell Atlantic must and will be in that market at the earliest possible opportunity or the firm will lose customers. The bottom line is that Bell Atlantic has more to gain from long-distance entry than any local telephone company in the country, so the firm is absolutely committed to local competition as the price to pay for cracking the long-distance market.

Thus, Bell Atlantic has done more than any other local operating company in the country to open its network to competitors. In addition to the 300 interconnection agreements, Bell Atlantic has been working long and hard with competitive local carriers, state commissions, and the FCC to fulfill the fourteen-point checklist—an extraordinary investment of technical, human, and financial resources. Bell Atlantic has not quite satisfied all the requirements in every area in its region but has almost done so in New York and expects to refile with the New York Public Service Commission by November 1997. When filing our applications for long-distance entry with the FCC early in 1998, Bell Atlantic intends to do so correctly because its filing is not exploratory.

That enormous undertaking involves every customer line, every central office, and an overwhelming majority of employees in the company. Bell Atlantic has spent close to $1 billion to open its network, has dedicated more than 1,000 people to that effort, and has created an entire business unit for the sole purpose of serving competitive local exchange carriers.

Bell Atlantic has developed constructive working relationships with a growing number of competitive carriers by processing tens of thousands of orders and providing those carriers with a level of service that enables them to compete for the local customer. In addition, Bell Atlantic has invested in the mechanized operating systems to process an even higher volume of orders in an orderly, timely way. In fact, in a massive test of those systems in New York, Bell Atlantic successfully processed five times the number of orders currently placed by competitive carriers on any given day, with an 84 percent flow-through rate and quick order confirmations.

How serious is Bell Atlantic about local competition? Consider the facts. The firm has real employees serving real competitive carriers by processing 1,000 to 2,000 orders in a single day in New York alone, can provision at least five times that number of orders with electronic ordering systems today—easily keeping ahead of even the most aggressive forecasts of customer demand—and knows how to increase that capacity quickly if need be. Spending close to $1 billion, Bell Atlantic has invested more in promoting competitive local entry in its territory than the biggest interexchange carriers have in entering that market.

Perhaps other Bell companies think that litigating—not implementing the act—is the best business strategy for them, but Bell Atlantic does not because it cannot afford to pursue that strategy. That does not, however, mean that Bell Atlantic will not go to court to protect its interests when any agency has overstepped its jurisdiction or gone beyond what the act requires. After all, the act says that states set the prices for interconnection, but under Reed Hundt's chairmanship the FCC is claiming that it sets those prices. Bell Atlantic cannot and will not overlook clear misinterpretations of the statute when they occur.

In any case, none of that slows Bell Atlantic's momentum in determining how to meet the guidelines so as to be able to file a compelling application for long-distance entry. Bell Atlantic is moving as fast as it can to speed that process and be the full-service provider that its customers want and need.

To review, local competition in Bell Atlantic territory is real. It is available through resale and unbundled elements. It is also facilities-based, with billions of dollars being invested in fiber and switches to carry local traffic. Local competition is here with a vengeance in short-haul toll, in data services, and in the large business marketplace, and it is picking up speed as prices for alternative technologies like wireless and cable reach parity with traditional phone service.

A short eighteen months since the Telecommunications Act was passed, there is no legal obstacle to competition in Bell Atlantic's local market. Today, any competitive local carrier could sell a full bundle of communications services to any home or business in its territory, either on its own facilities or by reselling Bell Atlantic's local loop and dialtone at the discounted rates that have been negotiated in all the firm's fourteen jurisdictions.

But the big question remains: Why are competitors not doing so? Or, to borrow from the simplistic language of the long-distance company ad campaigns: Why does John Q.

Public not have a choice of local phone company? To get to the motives behind that question, it is useful to distinguish between two kinds of competitors. In the first category are companies with an economic incentive to be in the market—entrepreneurial companies with smart business plans like Eastern Telelogic, Winstar, Teleport, Brooks Fiber, WorldCom, and countless others. After WorldCom's bid for MCI, it is hard to argue that those competitors do not have the scale or scope to be significant. They are world-class competitors with world-class facilities, and Bell Atlantic is doing business with them every day.

The second category is what I call the "phantom competitors"—notably the big interexchange companies that pretend that Bell Atlantic is preventing them from entering the market so that they can keep Bell Atlantic out of the long-distance market and blame the firm for their failure to develop working systems and processes. Unlike "real" competitors, "phantom" competitors make no serious attempt to sell local service, either on their own facilities or by reselling Bell Atlantic's. They send just enough "test" orders through the system to report hypothetical "problems" to regulators in the hope of delaying the entry of the only competitors that could change the price dynamics of the profitable long-distance cartel.

That is not rhetoric; it is fact. For example, of the more than 100,000 resale lines sold in the state of New York as of August 1997, less than 1 percent—a mere 700 lines—was sold by MCI. One small company in Chicago—USN Communications—processes more orders in one hour than AT&T has ordered since the Telecommunications Act was passed. The truth is that the leading long-distance companies lack the will, the leadership, and the strategy to compete in local service markets, and no amount of subsidized pricing schemes or other regulatory inducements will draw those reluctant gladiators into the ring. The best business strategy the interexchange carriers have come up with is to keep Bell Atlantic in regulatory quarantine, and the best way to do that is not to compete for residential customers at all. Therefore, further regulatory tinkering and evermore byzantine price schemes will only play into their hands. As long as they can exercise their de facto veto power over Bell Atlantic's entry into their markets, they have no economic incentive to enter, in any meaningful way, Bell Atlantic's.

In short, if a resident of the Bell Atlantic region has not been called by a competitor offering a choice of local service carrier, that is a result of a business decision on the part of the long-distance companies, not of any action or inaction on the part of Bell Atlantic. The only thing that will draw those behemoths into the ring is to let Bell Atlantic compete in the long-distance market. Every bit of market research suggests that customers want one-stop shopping when it comes to communications. When Bell Atlantic can offer a full bundle of services, its competitors will be forced to do the same.

Those who want real competition in communications should allow Bell Atlantic to enter the long-distance market. That will provide a wake-up call to long-distance carriers, push aside regulatory games and strategies, and let everyone know that the FCC will not protect AT&T and MCI from real competition forever.

Senator John McCain was right when he reminded his Senate colleagues that the Telecommunications Act was not designed to redistribute local market share to the long-distance companies. Rather, the stated purpose of the act is to "open all telecommunications markets to competition." The test of competition, Senator McCain has asserted, is whether local markets are "reasonably—not actually or perfectly—open." Bell Atlantic can take a page from Senator McCain's book in speaking reasonably about where the firm stands in implementing the Telecommunications Act.

Any reasonable examination of Bell Atlantic's actions in opening its local network would reach the following conclusions. First, Bell Atlantic has demonstrated its ability to work through the extraordinarily complex issues embodied in the competitive checklist. Second, the firm has demonstrated its willingness to deliver a level of service to competitive local exchange carriers that gives them what they need to compete and meets the requirements of some of the most demanding regulators in the country. Third, Bell Atlantic has addressed every legitimate issue posed by the FCC, state regulators, Congress, and competitive local providers. Finally, a healthy, competitive, facilities-based local service industry in Bell Atlantic territory continues to blossom.

In short, Bell Atlantic has answered the call of competition with action, not rhetoric. Bell Atlantic is making competition work in the marketplace and has put its facts on the table for all to examine. Now it is time for the rest of the industry to do the same. I call on the entire communications industry—as well as the new commissioners at the FCC and the state regulators throughout Bell Atlantic's fourteen jurisdictions—to turn off the smoke machines, cut out the negative advertising, and work in good faith to deliver what the American public has said it wants: true choice in all communications markets.

I cannot speak for the rest of the local exchange industry or for other telephone companies that may have different economic interests and different strategies. I can speak only for Bell Atlantic. Bell Atlantic has a tremendous opportunity to ensure that its customers are served by the world's best communications companies by fighting in the marketplace, not the courts, for the privilege of serving customers' communications needs.

Delivered October 7, 1997.

4

Abolish the FCC and Let Common Law Rule the Telecosm

Peter W. Huber

Herbert Hoover had a problem. As a Republican, he liked free markets and private property. But as an engineer, he wanted things nice and tidy. Markets are not tidy. He had to choose. He chose wrong. Hoover was secretary of commerce then, and the critical choice he had to make concerned a fascinating new industry called broadcasting.

Westinghouse had inaugurated the first radio station, KDKA, in Pittsburgh in 1920. AT&T put a station on the air in 1922. Hundreds of stations were popping up all over the country. And no one was regulating them at all. They registered with Hoover's Department of Commerce, in much the same way as ships registered. But that was it. Nobody stopped anybody from going on the air.

What happened when an AT&T station tried to broadcast on a frequency already used by Westinghouse? The companies went to court. Courts were just beginning to develop rights of private property in spectrum. They relied on familiar, common-law principles developed over the centuries in connection with real estate. But the common law is a messy business—as messy as the marketplace. Hoover, the engineer, wanted order.

Order had already come to telephones. In 1902 telephones had been messy, too. Half of all cities had two or more competing phone companies. They did not connect with each other. That was a nuisance.

In 1913 antitrust litigators hit on a solution. A consent decree forced AT&T to connect its long-distance network to independent phone companies. But by then it was already too late. People much like Herbert Hoover had taken charge. They solved the interconnection problem by abolishing competition. With only one phone company in town, it would not have to connect to anything but itself. That was tidy.

So by 1927 Hoover knew what he had to do. With the help of Congress, he nationalized the airwaves completely. He handed all the ether over to a brand-new Federal Radio Commission. From then on, private ownership of any part of the airwaves was strictly forbidden. Market forces were banished. Private users operated only under short-term licenses and under close federal supervision. The logic for that radical collectivism? Spectrum was too scarce to be owned privately.

And so in 1934—while Stalin, Hitler, and Mussolini were proving to the world the remarkable efficiency of national socialism, and while George Orwell was formulating his gloomy views about Big Brother—we in the United States folded all federal authority over both wireline and wireless communication into a new, superpowerful communications commission. We called it the Federal Communications Commission—the FCC. Germany got an FCC, too, even bigger and more efficient than ours. They called it the Bundespost. Joseph Goebbels loved it.

The FCC never got as bad as that, but it got bad enough. Our airwaves were about as free as a high-school newspaper. People lucky enough to get a broadcast license took care not to offend the commission. They wanted that license renewed.

And telephones? In 1954 the FCC invoked the full majesty and authority of the federal government to block sales of the Hush-a-Phone, a small metal cup that attached to the mouthpiece of a telephone to provide a bit of privacy and quiet in crowded offices. The FCC said it was a "foreign attachment" to the Bell network.

The years passed. And then suddenly Americans woke up one day, and whom did they find on their telephone network? Markus Hess, a German teenage hacker: Joseph Goebbels turned inside out, whose 1986 exploits were recounted in Clifford Stoll's *The Cuckoo's Egg.*

Using pseudonyms like Pengo, Zombie, and Frimp, Hess and others like him were bent on creating electronic anarchy, beginning with the phone companies. They attacked computers in Germany, Switzerland, France, and the United States. For a while they seemed able to go anywhere they liked. And they did it all over the phone wires.

What had happened between the time that the FCC required protective connecting arrangements to connect non-Bell equipment to the Bell network and the time that people with names like Zombie and Frimp were ambling down the international phone lines and into the most private recesses of Americans' computers? Two things. First, Bell Labs had invented dialtone for computers: an operating system called Unix, designed for computer networking. Second, in the 1970s the FCC had finally grown fed up with the silliness of regulating customer-provided equipment. So the commission began instructing Bell to connect with its most direct rivals: first, with the competition's phones, modems, and computers, next with long-distance competitors like MCI, and then with on-line servers—the building blocks of the Internet. Big pieces of telephony were being handed back to the competitive marketplace.

Consumers were startled. They had been living under rigid state control for a half century. They did not know a thing about free enterprise in telephony. They were like the Russians in 1992, trying to learn how to handle the free market. For a while, hustlers like Zombie and Frimp swept in and conned consumers.

Discipline was collapsing in the airwaves, too. The FCC had grown tired of running high school newspapers, so it pretty much told broadcasters to broadcast what they liked.

Americans got Jenny Jones and Rikki Lake—the Zombies and Frimps of the airwaves—in their living rooms. They were too stunned even to turn the dial.

Herbert Hoover gave us the state, and when the state began to crumble, Markus Hess gave Americans a taste of anarchy. It is time now to return to where we started: common law for the telecosm. That is the closest we shall ever come to utopia.

Telecosm

Until fairly recently, the telephone industry operated under laws that had their roots in the early twentieth century, a time when land, air, water, and energy all seemed abundant, and the telecosm seemed small and crowded, a place of scarcity, cartel, and monopoly—one that required strict rationing and tight, central control.

In the past decade, however, glass and silicon have amplified beyond all prior recognition people's power to communicate. Engineers double the capacity of the wires and the radios about every two years, again and again and again. New technology has replaced scarcity with abundance and cartels with competition.

The electronic web of connection that is now being woven among us all is a catalyst for change more powerful than Gutenberg's press or Goebbels's radio. Every constraint of the old order is crumbling. The limitless, anarchic possibilities of the telecosm—the universe of communications and computers—contrast sharply with the limits to growth people now encounter at every turn in the physical world.

In early 1996 Congress passed the most important piece of economic legislation of the twentieth century. The Telecommunications Act of 1996 runs some one hundred pages. The act's ostensible purpose is to open markets to competition and deregulate them. It may eventually have that effect. The process of deregulating, however, seems to require more regulation than ever. The FCC no longer aspires to immortality through its work. Like Woody Allen, it aspires to immortality through not dying. But it is time for a fundamental change; it is time for the Federal Communications Commission to go.

The Future and the Past

The telecosm is expanding faster than any other technocosm has ever done. It is the telephone unleashed, the personal computer connected, and the television brought down to human scale at last. Its capacity to carry information has expanded a millionfold in the past decade or two. The telecosm will expand another millionfold in our lifetime—or perhaps a billionfold. No one really knows. The only certainty is that the change will be enormous.

That change is characterized by a paradox: It is both fragmentation and convergence. The old integrated, centralized media are being broken apart. Terminals—dumb endpoints to the network—are giving way to "seminals"—nodes of equal rank that can process, switch, store, and retrieve information with a power that was once lodged exclusively in the massive switches and mainframe computers housed in fortified basements. That is the fragmentation. At the same time the functions of those nodes are coming together. In digital systems a bit is a bit, whether it represents a hiccup in a voice conversation, a digit in a stock quotation, or a pixel of light in a rerun of *I Love Lucy.* That is the convergence.

Then there is the law. Until 1996 most of telephony was viewed as a "natural monopoly." The high cost of a fixed plant, the steadily declining average cost of service, and the need for all customers to interconnect with one another made monopoly seem inevitable. The broadcast industry was viewed as a natural oligopoly. It depended on inherently "scarce" airwaves and was therefore populated by a small, government-appointed elite.

The FCC and comparable state-level commissions were established in the 1920s and 1930s to ration the scarcity and police the monopoly. The administrative structures, their statutory mandates, and the whole logic of commission control reflected the political attitudes of the New Deal. Markets did not work; government did. Competition was wasteful; central planning was efficient. Policymakers made a fateful choice: They rejected the marketplace and common law and embraced central planning and the commission.

The common law evolves from the bottom up. Private action comes first. Rules follow, when private conflicts arise and are brought to court. Commission law was to be top-down. A government corps of managers, lawyers, economists, and technicians would settle in at the FCC first. Private action would follow later, when authorized. Common law is created by the accretion of small rulings in discrete, crystallized controversies. Commission law would be published in elaborate statutes and 10,000-page rule books. While those were being written, the world would wait. Common law centers on contract and property—legal concepts that are themselves creations of the common law. Commission law would center on public edicts, licenses, and permits. Common law is developed and enforced largely by private litigants. Commission law would come to court only at the end of the process, when public prosecutors filed suits against private miscreants.

Common law would have suited the American ethic of governance far better, particularly in matters so directly related to free speech. But between 1927 and 1934, when the FCC was erected, the winds of history were blowing in the opposite direction. National socialism, right-wing or left, seemed efficient—the only workable approach to modern industrialism. Around the globe, people in power persuaded themselves that the technical complexities of broadcasting and the natural-monopoly economics of telephony had to be managed through centralized control. The night of totalitarian government, always said to be descending on America, came to earth only in Europe. But America was darkened by some of the same shadows. One was the FCC.

Once in place, the FCC grew and grew. Today, it has 2,200 full-time employees and a $200 million budget—more offices, more employees, and more money than at any other time in its history. As competition increases, monopolies fade, and the supposed scarcity of spectrum is engineered into vast abundance, the commission just gets bigger. An institution created to ration scarcity now thrives by brokering plenty. It is an Alice-in-Wonderland sort of world, in which the less reason the queen has to exist at all, the more corpulent and powerful she becomes.

For the next several years at least, the FCC will have the most important mission in Washington. Wireline and wireless telephony, broadcasting, cable, and significant aspects of network computing together generate some $200 billion in revenues a year. For better or worse, the FCC will profoundly influence how they all develop. And in so doing, it will exert a pivotal influence over the entire infrastructure of the information age and thus the economy, culture, and society of the twenty-first century.

The faster that power is dissipated, the better it will be for America.

Pricing the Commission

Had the commission permitted consumers to buy their own telephones in 1956, when it first considered the issue, instead of 1975, when it finally said yes, consumers would have saved over $3 billion. Had the commission authorized full long-distance competition when that was first proposed in 1968, rather than in 1978, consumers would have saved $16 billion. Had the commission authorized local phone companies to offer voice mail services in 1981 rather than in 1988, consumers would have been $6 billion richer. Had the commission authorized phone companies to begin competing directly against cable in 1984, cable consumers would have saved between $3 billion and $7 billion a year between 1983 and 1992.

Delays in licensing spectrum have imposed even higher costs. Unlike oil, gas, or timber, spectrum does not dissipate. If it just sits idle for a year, that is an irretrievable loss. If FCC licensing procedures for radio and television slowed the development of that industry by as little as a year, that delay cost the economy, over time, about $38 billion. Cellular telephony was conceived in 1947; AT&T began operating its first experimental system in 1962. But the FCC did not get around to issuing cellular licenses until 1983. Nationwide cellular service could have been in place at least a decade earlier; consumers lost at least $85 billion because it was not. Six years of FCC-imposed delay in the 1980s and early 1990s before licensing a next-generation wireless phone service cost consumers an additional $9 billion.

Delays aside, the commission limits what licensees can do in ways that sharply reduce the value of licenses once issued. If a single UHF television station in Los Angeles were free to shut down and transfer its spectrum to a third cellular provider, the public gain would be about $1 billion over eight years. Had the commission permitted FM subcarrier frequencies to be used for services other than broadcast services in 1948, instead of 1983, the economy would have gained nearly $2 billion dollars.

When the FCC is not issuing licenses, it writes rules. Commission rulemaking has degenerated into a second quagmire of false starts, delay, backtracking, and endless fine-tuning. For example, the commission has been equivocating for three decades over the rules to govern phone company provision of "enhanced services" like voice mail. During that period the commission has issued over 150 orders. It has made five trips to appellate courts to defend its rules. It took nearly twenty-five years before competition in the provision of customer premises equipment (like telephone handsets) was permitted, and the commission spent a comparable amount of time in a misguided attempt to protect broadcast from the "threat" of cable TV.

Throughout its existence the commission has also tinkered with broadcast program formats; those rules have served no demonstrably useful purpose at all. During the course of nearly forty years, the commission issued more than twenty orders in enforcing the Fairness Doctrine, only to decide in the end that telling broadcasters what to air did not serve the public interest and was unconstitutional to boot. The commission wasted three decades trying to dictate what kind of programming should or should not be carried on cable before largely abandoning the effort.

The commission likewise spent three decades enforcing its financial syndication rules—rules that governed economic relationships between broadcast networks and Hollywood studios. To that end the commission wrote fifteen major opinions and a lengthy staff report. Three times the commission had to defend its action before federal appellate courts. During much of that period, a comprehensive antitrust consent decree addressed precisely the same concerns. Since 1940, the FCC has issued a labyrinth of rules addressing how many broadcast licenses a single entity may own and limiting cross-ownership between broadcasters and other media. The rules keep changing. Yet antitrust laws enforced independently by the Department of Justice serve precisely the same function.

In one way or another, the FCC also attempts to regulate prices throughout almost every sector of the telecommunications industry. Price controls are rationalized as necessary to protect consumers. Time and again, however, FCC regulation has had precisely the opposite effect. Commission interference has maintained price rigidity and suppressed competition. In regulating telephony the commission for years allocated costs between state and federal jurisdictions in defiance of all norms of economic efficiency. For decades the FCC also did its utmost to protect free broadcasting from any for-pay alternatives. The commission allowed broadcast options such as pay-TV only after first subjecting them to years of crippling restrictions. In addition, cable was prevented from bidding popular programming away from broadcast TV. The commission has performed no better in its recent attempts to regulate cable television prices. Meanwhile, the FCC has delayed or blocked competition from satellite operators and phone companies.

The FCC often characterizes itself as guardian and promoter of universal service. Time and again, however, the commission's main strategy for promoting universal service has been to protect monopoly. For decades, the commission approved inflated long-distance rates to subsidize below-cost local telephone service. When it finally accepted competition and allowed prices to move closer to costs, telephone penetration increased everywhere. At various times, the commission has embraced AM radio, then FM radio, then television as the main beneficiary of its "universal broadcast" policies. The commission began by suppressing each new technology before it finally got around to promoting the technology. Only after decades of treating cable as an enemy of universal "free" television did the commission finally accept it as a new standard for "universal service."

All that teaches only what we already knew. Government itself is a monopoly and an imperfect one. It is a monopoly quite capable of imposing vast burdens on economic welfare. Private monopoly can do that too, of course. But the best possible environment is not one in which a private monopoly is held checked by the best efforts of a public one. It is one in which competition in the marketplace substitutes for both.

Deconstructing the Telecosm

The beginning of the end was cable television. Cable demonstrated that spectrum could be bottled and made abundant. Neither "broadcaster" nor "carrier," it is a capacious, flexible, constantly changing mix of both. Cable has already established that competition against broadcast is possible. There is no "scarcity" of television any more; wires offer limitless amounts of new spectrum in the confines of a metal duct. Now cable is moving into telephony. Meanwhile, by boosting the capacities of their wires, phone companies are poised to move into video; they already carry most of the Internet traffic, which is television in slow motion.

After almost two decades of trying to stifle cable, regulators cut it loose in 1984. In 1992 they tried to tether it again, then began cutting again in 1996. As regulators loosen their grip on cable, which should never have been regulated in the first place, the case for deregulating other media becomes compelling. Deregulation spurs competition, and competition forces still more deregulation. Thus, other wireline media are being destructured in cable's wake.

Telephone companies are finding their own avenues of escape from the commission, by invoking their First Amendment rights and exploiting logical contradictions that are now apparent in the Telecommunications Act. By bundling content and its transport, the phone companies can redefine their services as "cable." By narrowing service and to whom they offer it, they can redefine their services as "private" (not "common") carriage, and that too suspends most regulation.

The wireless universe is being deconstructed as well. Once dedicated largely to television, wireless is now the booming center of cellular telephony, direct broadcast satellite, wireless cable, and personal communications services. Too slowly, but inexorably nonetheless, spectrum is being privatized and dezoned. The new owners are using their wireless bandwidth to provide whatever service they like, to whichever customers they choose.

The fundamentals of deregulation are now clear. The concepts are simple. They can be implemented quickly.

First, throw open the markets. For wireless, that means privatizing the critical asset—spectrum—by giving it away or, better still, by selling it. For wires, that means letting anyone deploy new metal and glass alongside the old. Contrary to what Congress assumed for half a century, no commission is needed to protect against "wasteful duplication," "ruinous competition," or "inefficient deployment of resources." Markets take care of that.

Second, dezone the bandwidth. No government office should zone some bandwidth for pictures, some for voice, and some for data. The market can work that out far better than any central planner.

Will new robber barons then buy up all the wires, corner the spectrum, jack up prices, ruin service, and impoverish consumers? With the entire industry in ferment, and with the telecosm expanding at big-bang rates, those fears are utterly implausible. But in any event, the traditional antitrust laws will remain in place. For all practical purposes, antitrust law is common law. It addresses specific problems in courts, not commissions. It is decentralized, adaptable, and resilient. Inflexible commissions just get in the way. Indeed, for decades the FCC has legitimized telecommunications practices that antitrust courts would never have tolerated in the commission's absence.

Ironically, the commission can justify much of its current frenetic activity by blaming its predecessors. If the airwaves had not been nationalized in 1927, they would not have to be sold off today. If the FCC had not spent a half century protecting telephone monopolies, it would not have to dismantle them now. If the commission had not spent so long separating carriage from broadcast, broadcast from cable, and cable from carriage, it would not have to be desegregating those media today. If it had not worked so diligently to outlaw competitive entry back then, it would not have to labor so hard to promote it now.

"I broke it then, so I'll fix it now" has a certain logic to it, even if the confession of past sins is always much less emphatic than the promise of the future atonement. But the fixing somehow always seems to take as long as or longer than the breaking. And while the commission plans and plans for perfect competition, competition itself waits in the wings.

The telecosm would be vastly more competitive today if Congress had just stayed out of session in 1927, in 1934, in 1984, and again in 1992, if Congress had never created the Federal Radio Commission, never folded it into the FCC, never extended the commission's jurisdiction to cable, and never expanded the commission's powers over cable further still. The 1996 legislation guarantees that the commission will grow in size and influence while it uproots the anticompetitive vineyard planted and cultivated by its predecessors.

But the uprooting should be done quickly. Five years is time enough; ten would be excessively long. And then? Then the commission should shut its doors, once and for all, and never darken American liberty again.

The Marketplace of Ideas

In a competitive telecosm, the challenge will not be to get all the people onto the network; it will be to keep some of them off. The old problem of censorship has been solved; the new challenge is how to check indiscriminate communication that defames, infringes on copyrights, invades privacy, or incites pathological conduct.

Here again, FCC-crafted law has been a disaster. In pursuit of simple-minded broadcasting for the masses, the commission presumed that the means of transport were scarce and inseparable from the message delivered. A few broadcasters were therefore anointed to supply news and entertainment to everyone. Their favored positions and the ubiquity of their reach became the main excuse for controlling what they transmitted. Having collectivized the medium, the commission could easily justify central control of the message.

But scrambling technology is now readily available and widely used. Full encryption or more modest forms of electronic warning labels can be added to cable, satellite, and ordinary terrestrial broadcasts. Thereafter, the "casts" are no longer "broad." They are received only by those who want to receive them. No commission is needed anymore to protect unwitting recipients—the sensitive, delicate, prudish, or easily offended. Set-top receivers supply private censors that perform far better than the public ones.

With abundant bandwidth and the technology for consumers to choose what to receive, broadcasting no longer has the characteristics that persuaded courts to corrupt the First Amendment or degrade copyright law. Anyone can "cast" broadly by leasing capacity on a cable network, a phone line, or a satellite, and hundreds of programmers already do. Anyone can choose not to receive most of what is cast, and that is what most of us do. Debates about "fairness," "diversity," or "political access" on video channels are as ridiculous as talk of fairness would be in connection with conversation on telephone lines. Bandwidth, like seats on a plane, is available for hire to anyone. With the networks now at hand, commissions only get in the way of diversity. The fastest way to diversify is to decommission.

The remaining problems center on communications that harm third parties by defaming, misappropriating, invading privacy, or inciting violence. But here, too, management by commission hinders far more than it helps.

Libel laws can take care of fairness; anything beyond creates more unfairness, not less. Other common-law actions between private litigants can likewise control unlawful conduct incited by free speech at least as well as that persistent problem ever can be controlled in a free society. Pathological responses to free speech can never be eradicated, but they cannot be ignored either. The best approach is to deal with any of those problems after the fact rather than before, in front of courts, not commissions, at the instigation of private litigants, not public prosecutors.

The FCC created a mess of copyright law too. The problem began again with the commission edict that broadcasting was the business of providing exclusively "free" television to anyone who wanted to pluck it out of the air. That notion could not be squared with copyright laws, and when the plucking began, the courts just gave up on protecting broadcasters against piracy by cable. Congress responded with another abomination—a Copyright Commission—that seized the copyrights of broadcasters, collectivized their ownership, and then resold them at depressed prices to cable operators. Now courts are slowly rediscovering a private law of intellectual property in obscure corners of wiretapping statutes and the common law. Here, again, the fastest way to get better law is to get less FCC.

Finally, in the century of Hitler and Stalin, no one can seriously suppose that a federal commission is where one goes to protect the privacy of the common man. Central government authority is the problem, not the solution. Common-law actions enforced by private litigants can develop a real law of privacy, once commissions have been swept aside.

Common Law for the Telecosm

Who, then, will maintain order in all those areas when the FCC is gone? Private actors and private litigants, common-law courts, and the market. It is the commission that must go, not the rule of law.

We still need laws to defend the property rights of people who lay wires and build transmitters, to enforce contracts and carriage agreements, to defend the freedom to speak and to listen, and to protect copyright and privacy. Anarchy works no better in virtuality than in actuality. The question is not whether there will be rules of law, but whence they will come.

Commissions proclaim the "public interest, convenience, and necessity." They issue general edicts. They publish rules in the massive *Code of Federal Regulations*. Common law, by contrast, evolves from rulings handed down by many different judges in many different courtrooms. The good rules gain acceptance by the community at large, as people conform their conduct to rulings that make practical sense. In that kind of jurisprudence, constitutions and codes provide, at most, a broad, general mandate to develop the law by adjudication. They operate like the Bill of Rights or the Sherman Act.

Commission law has been tried. Not just in the telecosm, but in command-and-control economies around the globe. Like Communism, commission law has failed. It is rigid, slow, and—despite all the earnest expertise of bureaucrats—ignorant. Market forces, mediated by common law, elicit information faster and more reliably. Markets constantly probe new technology, try out new forms of supply, and assess demand with a determination, precision, and persistence that no commission can ever match. Property-centered, contract-centered, common-law markets allow people to get on with life first and to litigate later, if they have to. Most of the time they do not. Rules evolve spontaneously in the marketplace and are mostly accepted by common consent. Common-law courts just keep things tidy at the edges.

The one strength of commission law is that it reduces uncertainty all around, but only because the market must wait for a commission to invent a whole framework of law upfront. That often takes years, and the framework is always rigid and inadequate. In a universe where technology transforms itself every few months, where supply and demand grow apace, where new trillion-dollar economies can emerge from thin air in a decade or so, uncertainty is a sign of health and vigor. In a place like that, nothing except common law can keep up. The law must build itself the old-fashioned way, through action in the market first and reaction in the courts thereafter.

If that suggestion seems outlandish, it is only so because the FCC has been around so long that people can no longer imagine life without it. Once Henry VIII's licensing of printing presses had become routine, it would have seemed equally outlandish to suggest that such an unfamiliar, complicated, and important technology might be left to open markets and common-law courts. When the Interstate Commerce Commission was created in 1887, it seemed essential to proper management of railroads. But when the commission was abolished in early 1996, hardly anyone even noticed. We never did create a Federal Computer Commission. The computer industry has nonetheless developed interconnection rules and open systems, set reasonable prices, and delivered more hardware and more service to more people faster than any other industry in history.

Now, with the telecosm growing explosively all around us, with the cacophony of free markets already drowning out the reedy proclamations of a senescent FCC, the only outlandish proposal is that we should keep it.

Delivered October 21, 1997.

5

The Telecommunications Act and Its Infancy: A Review of the First Eighteen Months

Thomas J. Bliley, Jr.

Is the Telecommunications Act broken? The answer is no.

A little over a year ago, my daughter-in-law gave birth to my first grandson, Thomas Jerome Bliley IV. My critics on telecommunications legislation would have had him speaking Latin by now, on the basis of what they expected of the Telecommunications Act of 1996: "It's been eighteen months. Why am I not able to get cable through my phone jack? Where is my videophone? And why did my cable bill go up?"

Many in Washington agree with the premise of this series of assessments of the Telecommunications Act. People who say that Congress failed—failed to bring about competition fast enough, failed to be specific enough, or failed because the legislation was too specific—would have expected too much from T. J. as well. The only Latin that little fellow knows so far is *spitto, spittere, actui,* and *splash!* But he is walking, and he is starting to say a few words. All too soon now, he will run, jump, and play. He will grow into a man, and he will even learn Latin. It will all happen in time, just like competition in telecommunications.

The Telecommunications Act has set forces in motion that cannot be stopped. Those forces cannot be reversed—not by regulators, not by lawyers, and not by the courts. They are forces of competition that are as old and as vibrant as the marketplace itself.

In October of 1995, I sat in my office with Ray Smith, the chairman of Bell Atlantic. It was the day I told him my goals for telecommunications reform. I picked up the telephone and listened to the dialtone. "Just what I thought," I told him. "Bell Atlantic." I said I wanted a day when I would pick up that phone and have a choice in telephone service—and not just when I called in New York or Chicago, but when I called in Richmond or Norfolk.

At that time, hardly any interconnection agreements existed. Today, we have more than 1,200—more than 280 involving Ray Smith's Bell Atlantic alone, and his company has assigned more than 14 million phone numbers to those competitors.

In October of 1995, competition was illegal in local telephone service in some states and practically impossible in all. In 1996 the competitive local exchange carriers earned more than $1.8 billion—and that does not include the major long-distance companies. In 1998 those revenues were expected to hit the $5 billion mark, an estimate that does not include the hundreds of new carriers just getting started.

Congressional critics—the ones who think T. J. ought to be reading Cicero by now—say that things are not moving fast enough. What they are not telling you is that competition involves every one of the 160 million telephone lines in the United States. Every carrier must be able to handle a local call by using the facilities of any carrier the caller chooses. That is a big job, and so is the building of the databases so that customers can keep the same phone number when they change carriers. Yet in Washington, D.C., regulators, along with politicians who cannot find a lost Social Security check, say that Congress is not moving fast enough.

Just like little T. J., however, the competitive marketplace is taking its first steps. The experts tell us that local lines in service for the biggest competitive local exchange carriers will have increased fivefold by the fourth quarter of 1998. Those carriers will have increased their payroll by as much as 45 percent; indeed, at year-end 1997 competitive local carriers employed 12,000 Americans. The biggest competitive local carriers have attracted more than $12 billion in capital investment since passage of the Telecommunications Act, and the amount of fiber they are laying is doubling every two years.

It took MCI almost thirty years to get a sizable share of AT&T's market in long-distance service, from 1963, when it first went to Washington, where it had to ask the FCC for authorization to compete, until the mid-1990s. Some analysts say that the regional Bell operating companies will lose 18 percent of their market share for residential services by 2001, at which time the new competitive local exchange carriers will have 23 percent of the market for business services. That change will take only four years rather than decades.

Today, more than a million Americans earn their living making telecommunications equipment: the switches, the phones, the lines, and the transmission equipment, all of which are to competition what supply lines are to an army. Factory sales of telecommunications equipment were up 15 percent in just the first year of the act, and telephone company spending on public networks hit the $10 billion mark, more than six times the rate of increase just one year before.

The wireless industry is hiring 2,000 Americans every month—the result of the doubling of investment in wireless since 1994. Today, 50 million Americans have cellular service; by 2000 that number will reach 60 million. What draws those new customers are prices that are half what they were just ten years ago. Since APC started Sprint Spectrum in the Washington, D.C., area, rates for cellular service have fallen by as much as 55 percent.

Some criticize the Telecommunications Act from other perspectives. For instance, some view the mergers of

Bell Atlantic and NYNEX or of SBC and Pacific Telesis as moves that eliminate potential competitors. But Bell Atlantic's merger can only improve NYNEX service, and most of the mergers that have taken place since passage of the Telecommunications Act would have happened anyway.

In fact, the Telecommunications Act ensures competition, even without the huge concentrations of capital that those mergers represent to critics of Congress. The act created three ways to enter the market: a firm may build its own facilities; buy unbundled elements from incumbent firms; or resell incumbent firms' services.

With respect to complaints about rising cable television rates, the Telecommunications Act of 1996 does not deregulate those rates until 1999—and then only for the biggest cable companies and only on the upper tier of service, not basic service. The cable rate increases that we have already seen have come about as a result of the Cable Act of 1992, legislation passed by a Democratic Congress and vetoed by George Bush.

The best regulator of prices is the marketplace, not a regulator, and that is what is happening in video services, thanks to the Telecommunications Act. Currently, 10 percent of the households in the market have a direct broadcast satellite dish on their roofs. Consumers tired of cable rates and choices have alternative service options. In 1995 it was illegal for telephone companies to offer cable within their own regions. In 1997 Ameritech had thirty-eight franchises and was planning more.

We are now at a pivotal time in implementing the Telecommunications Act. The new FCC commissioners and chairman should move forward steadily but surely to implement the act. They should put faith in competition, the hallmark of the act. They should stake out a role for themselves, and for the commission, that is in harmony with the brave new world of telecommunications in the next century.

That is the true legacy of the Telecommunications Act, not the day-to-day squabbling of what I call the Washington legal-industrial establishment. The new commissioners must be men and women of vision, not ciphers; Washington already has plenty of those. The new commission must understand that the act is not about who is suing whom or who is filing what petition with the bureaucracy. That is the small stuff, important only for the moment, and even then only in Washington.

America's legacy will be the innovations that are unleashed by a whole new range of investments—innovations in fields, as I said when we wrote the act, that one cannot even imagine today. Those innovations include the one being worked on by United Utilities and Northern Telecom that could turn every household electrical outlet into a modem capable of receiving the Internet, and exciting new developments like real-time video and Internet broadcasting.

Those innovations are the real products of the act. Its true disciples are not the armies of Washington lawyers, but rather innovators like Mid-River, a lean and hungry company that has just signed up the majority of the families in one Montana town for local, long-distance, and cable services—while the incumbents were busy in court. That firm's true disciples are companies like RCN in New York City, which is beating up NYNEX and Time Warner in one shot by giving their new customers a 10 percent break in monthly rates for telephone and cable. Those are the true disciples, and that is what the Telecommunications Act is all about. Is it broken? Hardly. To even ask the question is like looking at a fine athlete, a champion runner, for instance, and saying, "Yes, but could she do that at eighteen months?"

Delivered October 28, 1997.

6

The Race for Local Competition: A Long-Distance Run, Not a Sprint

Joel I. Klein

That the question whether the Telecommunications Act of 1996 is broken, and, if so, how to fix it is before us suggests to me that many thought that the journey from local monopoly to competitive markets would be a quick and smooth one. The truth, however, is that no one who fully understands the economics or technical aspects of the telecommunications industry would have predicted that local telephone competition would blossom overnight—or even fully mature within two years of the act's passage. To be sure, during its inception, as well as during the maneuvering to get the legislation enacted, we did hear rosy predictions about cable telephony and successful forays by the long-distance companies into local telephone service, but those predictions were clearly self-serving and overly optimistic. Not surprisingly, such predictions led to unrealistic expectations and, consequently, to the disappointment of those who expected that consumers would quickly see the benefits of competition.

To understand why the act sets the right course, we must first get beyond the unrealistic expectations about how competition would unfold. We can then examine the basic steps necessary to move from regulated monopoly to competitive markets. Finally, once we understand the nature of such a market-opening process, we can assess where things are, what progress has been made, and what is likely to lie ahead.

It is fairly clear that the disappointments about the progress thus far do not reflect any inherent or irremediable flaws in the act, but rather, the unrealistic expectations set during the act's passage and, possibly, the understandable eagerness of our political system to cut short a deregulatory course that is designed for long-distance runners, not sprinters. My view is that the Telecommunications Act maps out a fundamentally sound approach, that we are making progress, albeit less than we all would like, and that we should stick to that course so that we can all benefit from the changes that will continue to take place as we bring competition to the local telephone market.

In saying that the act points in the right direction, I do not mean to imply that time alone will lead to full-scale competition. Implementing the act is a difficult task: much money and many entrenched interests are at stake, and tricky technical, technological, and legal issues remain to resolve. Moreover, realizing the act's vision of a competitive marketplace will necessitate a transition from a system containing implicit subsidies for incumbents to one providing only explicit, competitively neutral subsidies. We must, however, go forward. The old system of regulated monopoly was unsustainable: we could not maintain an inefficient, bifurcated market for local and long-distance services and its cross-subsidization of different services. Those who understand the industry know that attempting to do so not only would have undercut the movement toward the offering of one-stop shopping for telecommunications services, but would have involved the gradual "cherry picking," through bypass opportunities, of profitable local and exchange access customers, thereby resulting in a system of cross-subsidization that would be not only inefficient but ultimately in shambles.

The Telecommunications Act in Perspective

Before outlining some of the steps involved in facilitating competition in the local exchange segment of the industry, I look at the development of competition that eventually permitted deregulation of the long-distance market.

In 1974 the U.S. Department of Justice concluded that microwave technology allowed firms like MCI to compete against AT&T in the long-distance market but that AT&T was using its control over local exchanges to stall competition in that sector. Even after the department obtained divestiture relief through the 1982 modification of the consent decree entered against Western Electric and AT&T in 1956, twelve more years passed before sufficient competition took hold so that AT&T's long-distance prices were deregulated. During those twelve years, the Department of Justice, the Federal Communications Commission, and state public utility commissions all labored to restructure the regulatory framework to facilitate competition where regulated monopoly had previously reigned. Two particularly challenging undertakings in that regard were the implementation of the equal-access portions of the divestiture decree—which promised all long-distance carriers parity in access to the local telephone network—and the development of a complex system of charges to bill for such access.

As a result of the efforts to open up the long-distance market, competitors have increasingly entered that market, and prices have fallen dramatically. On average, consumers now pay 60 percent less, adjusted for inflation, than they did when AT&T was broken up in 1984. Consumers today have a choice of scores of long-distance providers, each with different options and offerings. And AT&T has been forced to compete for its customers, losing 40 percent of its market share since 1980, now even offering many consumers dime-a-minute long-distance phone calls.

Bringing competition to the local market will also take time —not twenty-one or even twelve years—but time. Indeed, in some respects, that undertaking is likely to be even more challenging than bringing competition to the long-distance market. By divesting the local operating companies from AT&T's long-distance operations, the AT&T divestiture decree removed the ability of the integrated Bell

System to use its control of essential inputs to discriminate in favor of itself vis-à-vis its competitors. And the development of alternative long-distance networks by new competitors proved to be reasonably affordable. By contrast, the creation of new networks and the risk of discrimination in their absence are at the heart of the challenge to bring competition to local telephony. The simple fact is that no one can be certain that, as in the long-distance market, we shall see the development of fully redundant local networks that could simply bypass the current incumbent monopolists. Consequently, the 1996 act properly charts a very different course from that adopted in the AT&T divestiture decree: it uses regulatory, contractual, and antitrust oversight to ensure that local companies share their essential inputs with their would-be competitors. The theory behind that approach is that once incumbents and new entrants have instituted the arrangements necessary for sharing essential facilities, those arrangements will become more and more regularized as well as legally enforceable.

At least two things make the type of sharing arrangements envisioned by the Telecommunications Act difficult to implement. First, shared inputs must be priced; and those prices can determine the relative success of new entrants vis-à-vis an existing incumbent monopolist. Not surprisingly, people are going to fight over those numbers. And each side understandably wants a margin of error to protect itself. Furthermore, as is always the case with price regulation, by definition, no market-based solution exists, so government agencies must serve as an imperfect substitute for the market. That inevitably adds a dimension of politics—real or perceived—to the process. The fight about pricing is what most of the current litigation is all about.

Second, even in circumstances where prices are not an obstacle, developing the relevant rules of the road, as well as the systems necessary for implementing those rules, often proves difficult in practice. Just as it took time to work out the handoff from cellular to local loops, so too will the handoff from a local incumbent to a new entrant take time—especially when that incumbent is being told to cooperate with an entity taking away one of its customers.

To deal with those matters, the Telecommunications Act adopted a two-part strategy: the sticks of enforceable obligations and the carrots of in-region, long-distance entry. As the act conceives it, the Bell operating companies must offer new entrants an equal opportunity to compete for local telephone service customers. Once they have achieved that goal, the act allows for Bell operating companies to enter the long-distance markets in their respective regions. In essence, then, the act envisions that the local and long-distance companies will enter one another's markets and offer new, improved, and less-expensive services, including bundled offerings of local and long-distance services, to consumers.

Obviously, the ability to implement the act will depend, at least in part, on the efficacy of the sticks and carrots, so I briefly consider each. With respect to the sticks, there are real questions. The act itself calls for no actual penalties for noncompliance and, at least in the Eighth Circuit's view, the FCC may have only limited—if any—authority to fine companies for failure to comply. While the state public utility commissions may be able to help, and some state legislatures—in Illinois, for example—have already adopted regulatory schemes to police compliance, most states have not instituted such schemes. Of course, the Telecommunications Act makes clear that section 2 of the Sherman Act still applies to the industry, and, generally speaking, that statute provides effective remedies to deter abuses of market power, with treble damages where appropriate. Nevertheless, as Congress wisely recognized, antitrust remedies are not well suited to serve as the first-line method for opening local markets. That fact of life, in combination with the uncertainty about how the statutory obligations are to be enforced, means that the carrot of long-distance entry appears to be the primary means of ensuring Bell operating company compliance with the act's market-opening provisions.

In that regard, it is apparent that different Bell operating companies have different appetites for that carrot, depending on each one's particular assessment of its own best interests. The Justice Department's experience has been that, while the Bell operating companies are obviously interested in in-region, long-distance entry, most are still trying to decide whether it is worth the price. By "price," I mean what the Justice Department and the FCC are requiring in the way of statutory compliance. To further complicate the analysis, several Bell operating companies appear to be weighing the likelihood that, despite the delay and uncertainty attendant to judicial proceedings, they will get a better deal from the courts than from the FCC and the Justice Department. My own sense of that assessment is that, while we shall continue to observe a range of responses, in the end, the Bell operating companies will conclude that the loss of desirable customers to new entrants, coupled with the appeal of becoming a one-stop shopping company, will ultimately lead to compliance with section 271 of the Telecommunications Act.

Opening Up the Local Market

Moving away from the big picture, I turn to some of the sticky details that need to be worked out in the wonderful period known as the "meantime." In particular, I want to consider how uncertainty—both in terms of what entrants can expect and in terms of what the law requires of incumbents—delays competition. In that regard, I focus on three issues: the development of the prices for the wholesale inputs; the legal disputes that have arisen in implementing the act's requirements concerning the combining of individual elements unbundled from an incumbent's network; and the technical development of the systems necessary to support competition in the local market.

On the pricing issue, a good bit of rhetorical crossfire has occurred in the regulatory and legal battles over the rates for leasing access and interconnection to an incumbent's network, as well as for purchasing resold services. At the end of that process—with whatever roles the state commissions, the federal courts, and the FCC will play—it will be necessary to put in place prices that will enable efficient competition to move forward. In the Justice Department's recent filing on BellSouth's section 271 application for South Carolina, the department explained that its statutory charge under section 271 to conduct a competitive assessment of the merits of any section 271 application allows it to evaluate, at a minimum, whether the prices in a given state are the product of a procompetitive methodology that was applied in a reasoned manner and will remain in place so that new entrants can make investment decisions with some confidence about the future pricing of essential inputs. I should emphasize that, under its competitive analysis, the department has insisted on a consistent set of procompetitive pricing principles, not on any single pricing formula. Thus, the department would not accept, for example, that backward-looking costs—such as embedded, uneconomic (or "stranded") costs or universal service subsidies—have a place in a wholesale pricing structure designed to foster efficient entry.

The second important issue in implementing the act is translating some of the broad statutory terms into specific legal rules. Although that process is moving forward in different contexts, one important issue addressed in the Justice Department's South Carolina evaluation relates to the legal machinations that have been going on about how an incumbent is supposed to allow an entrant to combine the individual unbundled elements of the incumbent's telephone network. Section 251(c)(3) of the act states that individual elements unbundled from an incumbent's network must be offered in a way that allows the requesting carrier to combine them to provide service. The FCC's August 1996 local competition order explained that the provision meant that new entrants were entitled to lease all the individual elements to provide service, even if those entrants were not using any of their own facilities. The Eighth Circuit upheld that regulation but vacated a second commission regulation requiring that incumbents provide new entrants with access to the already connected elements of their own networks.

Given the recent rulings on the "rebundling" issue, the Justice Department faces the uncomfortable and inefficient situation in which incumbents are allowed to do some work—uncombine elements already in place—so that new entrants can do more work—by recombining them. The department's evaluation of BellSouth's recent section 271 application explained that, because the company had failed to outline the specific process and procedures by which new entrants could do such recombining, it had failed to implement the mandate of section 251(c)(3). In the meantime, with litigation changing the regulatory landscape, the department has yet to determine what sorts of arrangements would provide reasonable access in that regard. The department has resisted any rush to judgment on that issue before learning from the companies themselves what solutions make sense given the need to implement section 251(c)(3).

The third issue in implementing provisions of the act is the challenge of instituting new technical arrangements that will enable competition to take root in local telephone markets. For simplicity, I use the term *wholesale support systems* to encompass the range of functionalities and operational systems that an incumbent must make available to a new entrant so that the entrant can switch over and serve customers previously served by the incumbent. Operations support systems are, as the FCC explained in its local competition order, important checklist items, as well as a means for making resold services and access to unbundled elements available in a meaningful manner.

As the Justice Department has explained in the three section 271 evaluations it has thus far filed, effective wholesale systems are essential to making meaningful competition possible. Obviously, a new entrant is at the mercy of an incumbent provider when it asks the incumbent to switch over a customer or to suppress the incumbent's billing systems so that a customer is not receiving bills from both a new entrant and the incumbent. With respect to resale, an incumbent's wholesale systems are even more critical because, unless the incumbent is processing orders in a timely fashion, a customer may be forced to wait for several weeks or even months before being switched over to a competitor, depending on the backlog of orders at the incumbent's headquarters. The rub there is that if service is not connected in a timely fashion, or if there are other operational problems, the customer will hold the new entrant—not the delinquent incumbent—responsible for the failure. Consequently, new entrants may tend to delay gearing up their operations until they gain a level of confidence in an incumbent's systems.

In addition to getting the right systems in place, the Justice Department needs to ensure that those systems remain in place after section 271 authority has been granted. As the department's evaluations indicate, it intends to accomplish that by insisting on appropriate performance measurements as a condition of entry. Thus, once a Bell operating company is reporting satisfactory levels of performance and has committed to performance standards, the department can support Bell entry with the knowledge that a "benchmark" of satisfactory performance has been set, thereby enabling postentry remedies—contractual, regulatory, and antitrust—to guard against any backsliding on the Bell operating company's earlier wholesale performance. With the right measures in place—the recent filing in South Carolina explains which ones the department considers important—postentry measures, such as halting future long-distance marketing authority, can serve as an effective incentive for a Bell operating company to maintain its levels of performance.

Those three areas are certainly not the only ones critical to competition, but they will need to be resolved effectively if large-scale competition is to move forward. They indicate why the market-opening process is not for sprinters. In addition, difficult problems often exist that make competition for local residential customers—especially outside major cities—a particularly elusive goal.

Competition for Residential Customers

Several considerations other than the obstacles to entry may help to explain the especially slow pace of residential competition: the relative costs of serving different classes of customers (for example, urban versus rural and business versus residential); our present system for supporting universal service; and delays in rolling out new technologies.

Serving business customers will often be more attractive to competitors because businesses generally use multiple lines and make more frequent and more expensive calls than residential customers. In addition, under our present system for ensuring affordable telephone service, certain services are priced at above-cost rates to subsidize other services, often leading to a situation where large business customers are far more attractive to competitors. Some states, for example, allow an incumbent to charge above-cost rates for multiple business lines to subsidize—albeit, implicitly—the incumbent's rates for basic dialtone service. To be sure, the act calls for those implicit subsidies to be made explicit and available to entrants seeking to serve high-cost areas, but until that transition occurs, entrants will be left with the prospect of competing to serve certain residential customers at below-cost rates—at least if they intend to serve those customers via unbundled elements. Of course, if new technological advances such as a wireless local loop or cable telephony enable new entrants to underprice an incumbent's subsidized residential service, that ability would facilitate residential local entry, but those advances do not appear to be imminent.

Of the factors that I just mentioned as bearing particularly on the pace of residential competition, only one is completely in the control of government: the system for ensuring universal service. As competition in urban areas and for business customers takes root, more and more fiber will be laid and more arrangements will be made for competitive offerings that will also benefit residential customers. As technology advances, the cost of delivering service may decrease considerably, thereby bringing residential customers new products at better prices. But, even as such developments take place, residential customers still will not get the full benefit of competition if policymakers continue to rely on a system of implicit, rather than explicit, subsidies that make at least some residential customers unattractive to competitive carriers.

In South Carolina, ACSI, a competitive local exchange company, reports that it must pay $19.45 per month for an unbundled loop and the associated charges—in addition to its other capital and operations costs—to serve a residential customer who presently pays $19.95 per month for basic dialtone. Perhaps ACSI could profitably serve such residential customers if they use vertical or enhanced features such as call waiting or voice mail or if they spend money on long-distance calls. But the firm certainly will not choose to market to customers who use only basic dialtone.

On the other hand, under an explicit system for subsidizing universal service, a customer would pay the affordable rate, say $19.95 per month in South Carolina, and a subsidy would be given to whichever carrier serves that customer. If the subsidy were explicit and available to all carriers, companies would find it profitable to go after all customers and would be able to compete with incumbents. The FCC and the states are now trying to make implicit subsidies explicit. Calculating what those subsidies might be in different areas is difficult and highly charged. But once an explicit, competitively neutral system for ensuring affordable universal service is put in place, it will provide an important incentive for competitors to go after the mass market. Until that is accomplished, however, we must candidly acknowledge that our present system of implicit universal service subsidies is a real impediment to full-scale residential competition.

Signs of What Is Yet to Come

Although there are problems in establishing local competition, we should be mindful of the tangible progress that has occurred thus far and of the auspicious signs for what lies ahead. According to Ameritech, competitive local carriers are now serving over 400,000 customers in its region, with more than 3,000 customers switching to one of Ameritech's competitors each day. Other regions are starting to see similar results. Just recently, Teleport, a relatively small new local entrant, was awarded a contract to serve the telecommunications needs of all city government offices in San Diego. That arrangement allowed the city to reduce its expenses by approximately 20 percent. Equally encouraging is the flow of capital being raised by companies gearing up to offering local telephony: in the first half of 1997 alone, competitive local exchange carriers raised over $6 billion.

In the future we shall see a greater convergence among utilities, cable, and telephone companies, bringing new and better services to consumers. In states such as Massachusetts, Texas, California, and New York and in the District of Columbia, cable and utility companies are employing excess capacity on their own, often underutilized lines to provide competitive local phone service. In Massachusetts, Boston Edison, an investor-owned electric utility company, through a partnership with RCN Telecom Services, is providing local, long-distance, data, and video services as part of a bundled service offering to approximately 4,000 subscribers. In the summer of 1997, CSW/ISG ChoiceCom—a partnership between an electric utility and a competitive local exchange company—reported that it had connected its first local service customers in Austin and Corpus Christi, Texas. At the same time, Cox Communications, a cable television company, launched a large-scale residential offering of digital telephone service in Orange County, California, to 1,500 households. Cox planned to be able to offer its phone services to 265,000 households in its service area by the end of 1998. In New York, Cablevision is rolling out—initially on a limited basis—its hybrid fiber coaxial technology to provide cable, telephone, and Internet access to customers on Long Island. In October 1997, Pepco announced plans to offer a high-speed, fiber-optic network that will provide cable, phone, and Internet service to its customers in Washington, D.C.

For those who want to look a little further into the future, the progress of competition in wireless services provides reason for optimism—and, potentially, even a full-scale competitor to local wireline service. In Texas, for example, PrimeCo, a wireless provider, charges a flat rate of ten cents a minute for intrastate calls, calls for which the local Bell company is allowed to charge up to thirty-six cents a minute. According to the *Wall Street Journal,* AT&T is developing digital wireless technology that could cost consumers as little as $10 per month for unlimited local calling. With the entry of additional wireless providers into many local markets, many cities, including Washington, D.C., now have five different wireless providers. One does not have to be an economist to recognize that good things start to happen with so many companies competing in a market.

Conclusion

In the midst of the overheated rhetoric by the industry participants, as well as the unmet expectations about what was promised during the debates over the act, some players would appear to be more interested in pointing fingers than in solving the problems necessary to make the act work. In my view, with the new technology on the horizon and the attractiveness of serving all of a customer's telecommunications needs—be they cable, wireless, local, long-distance, paging, Internet access, what have you—our old communications laws premised on regulated monopolies threatened to impede real progress in the industry. The 1996 act represented only the first—though a very bold—step with much hard work left to be done. But that work should be done. The Justice Department will continue working with the states, which have played a vital role in the process, as well as with the FCC, which now has new leadership to continue the march through the difficult terrain. At this point in time, when we are in the midst of the process of taking on the difficult tasks along the way, it is not always easy to keep an eye on the big picture, but I am confident that our journey to competition, though not easy, will be well worth the work.

Delivered November 5, 1997.

7

Damn the Torpedoes—Full Competition Ahead!

William P. Barr

Has the Telecommunications Act been a failure? I do not think so, and I believe that those who do say so have reached that conclusion prematurely. As Commissioner Susan Ness says, those people who talk about when competition will start are like children who ask, "Are we there yet?" before their parents get out of the driveway. That is a very apt description; it is too soon to be castigating the act. But the intentions of the act have been frustrated by the actions of regulators charged with implementing it, and I examine those actions.

Like any law, the Telecommunications Act is not perfect. It is a product of many compromises. I do not like many parts of the act, and I disagree with some of its provisions. But on balance, the act took a relatively sound approach to opening up the telecommunications markets to competition. As I view it, the act really meant to achieve two tasks.

The first was to set rules that would allow and facilitate entry into local markets. There, the act took a number of important steps. First, it eliminated the legal barriers to entry. It eliminated the franchise system so that any player wanting to come in, put in facilities, and enter can do so.

Second, through the interconnection requirement, the Telecommunications Act did away with the requirement of ubiquity. No longer does a firm have to build an entirely redundant system with a wire into every house to compete in the local market. A firm can now buy ubiquity through interconnection, and that means that the firm can tailor and target its network and serve consumers selectively. Those are very important elements for easing entry into the market.

Third, the act provides for resale. The argument, of course, of many of the would-be new entrants was, "We have to build up scale before we can make an investment." So the act wisely provides a resale regime, whereby new entrants can build up their customer bases before they actually have to start investing in facilities to serve them.

Finally, the act sets forth unbundling rules intended to permit the gradual deployment of facilities. A firm can build up its scale and enter in a staged manner, incrementally as it adds facilities. The firm does not have to put in the entire network. It can rent pieces that it does not have.

That is a sensible regime, as laid out in the statute.

The second of the overarching tasks of the statute was to eliminate the old way of funding the universal service system—that is, the cross-subsidies embedded in retail prices that required an exclusive franchise for the firm. Indeed, that is the fundamental task in the statute: the sine qua non of competition in the local market is to eliminate the old system of funding universal service and to come up with a new system. It is meaningless to talk about competition and opening up markets to competition unless one deals first with the gross pricing distortions that have developed over time to provide universal service.

Thus, we must recognize that the past regulation of the telephone industry was not simply directed at controlling a monopolist and ensuring that it did not gouge consumers with its prices. It was a decision by the body politic to pursue a social program, a program of redistributing income through the provision of telephone services. As that system grew, it fostered affordable rates for residential and rural customers that, in practice, have become below-cost rates for a large portion of the consuming public. To subsidize those rates, regulators relied on business customers, those who use long-distance telephony extensively, and on those who have many vertical and discretionary services on their phones.

Universal service is widely supported, even by the phone companies, but it is a social program that only a monopoly could maintain. The Federal Communications Commission's attempt to graft competition onto the market has a serious consequence. It turns the subsidy that used to keep Aunt Tillie's bill low and make her phone service affordable into an inducement for an inefficient new entrant to join the market and pocket the subsidy. The existence of more players in the marketplace is a result of regulatory rigging, not competition. Unless someone can serve Aunt Tillie and make a profit, the benefits of competition will not flow to her and her fellow residential customers.

Chairman Kennard said that the purpose of the act is to ensure that the benefits of competition reach all citizens of the United States. My plea to him and to the entire regulatory community is to straighten out universal service. Policy must ensure adequate support—not a low-balling of support.

Why have the regulators not come to grips with the issue of providing cross-subsidies to achieve universal service—the real key to opening up a market to competition? I think, for some, maintaining that cross-subsidization scheme is a deliberate effort to create the illusion of competition and to force, through regulatory fiat, the reallocation of market share. For others, maintaining that subsidy is a political calculation—the terror that if Aunt Tillie's phone bill goes up $4.00 a month, people will riot in the streets. The politicians, not the local telephone companies, must make that call. To what degree are politicians willing to subsidize low rates? What should those rates be? Whatever the answers, policymakers must ensure that a neutral and sufficient mechanism exists to pay that subsidy. Some states—California, for example—have already rebalanced consumer prices, and I have not heard of a revolution over that.

But policymakers have made few such forthright efforts. We see instead, for example, access charges. One might think that, just as a starting point, people would be willing to raise the subscriber-line fee to reflect inflation. Instead, however, the FCC is going to impose a new charge on services subject to competition and compete the universal-service subsidy away. Such action simply does not make any sense.

Another bizarre example is the FCC's universal service order, which says: "We're going to measure how much subsidy is necessary, but we're going to do it based on average revenues per line." That is simply the old system wherein the commission made sure that average revenues equaled average costs. To decide now that the commission is going to measure how much subsidy is needed by looking at average revenues keeps those old, implicit subsidies in place.

As I said earlier, the second major task was to set the rules to open local markets. Unfortunately, a lot of attention has centered on the rules for unbundled network elements and wholesale services. Let us not forget that to compete entrants have ways other than piggybacking on an incumbent's network. New entrants can put facilities into the ground, for example, as many competing local exchange carriers have done for a long time.

But a debate about the unbundling rules relates to pricing an incumbent's inputs when they are used by new entrants. What has happened there is another gross distortion of the purposes of the Telecommunications Act that is inexplicable from the standpoint of public policy.

It would be one thing to leave the current system of universal service in place and allow entrants to take advantage of the incumbent's subsidy system without having to pay those implicit subsidies. But the unbundling rules for network elements and the pricing rules related to those elements actually facilitate and exacerbate the problems of cherry-picking and cream-skimming.

GTE has three objections to TELRIC (total element long-run incremental cost) pricing, the pricing system used for unbundled network elements. In particular, GTE objects to forward-looking pricing based on a hypothetical network. Pricing should be based on actual networks.

Second, much confusion over historical costs exists. As long as some outboard mechanisms exist for dealing with the issue of stranded investment and the implicit subsidy, GTE has no objection to using forward-looking costs as the basis for pricing the unbundled network elements.

Third, an input sold to a new entrant at too low a price preempts real competition. The new entrant will use that below-cost element, or input, and not provide its own input at all. But an input with too high a price will invite competition. Competitors will put in a switch at a lower price, and the market will adjust the price downward.

I am afraid that regulators have gone too far toward pricing on the low side—50 to 60 percent discounts from actual costs. So they have preempted competition and eliminated the prospect of firms' entering and making investments of their own. One cannot simply reconstruct an unbundled local telephone system for half of what it cost to build. If AT&T could use its much-vaunted 51 percent discount in Pennsylvania to construct a local phone system there, it would—and it would kill Bell Atlantic. But, of course, AT&T cannot do so, so it looks for artificial price and the so-called unbundled network elements as means of entry.

Regulators need to hear the following message from incumbents:

> If you are going to make us sell or lease our network to a new entrant and you have not yet taken care of universal service—if you continue to distort our retail rates with built-in subsidies—then you have to reflect that in the pricing of unbundled network elements.

One approach is to let incumbents reduce their prices, bleed out those subsidies, and come up with a competitively neutral way of supporting universal service. But if regulators reject that alternative, the incumbents' message is simple:

> If you still expect our business rates to carry substantial subsidies, then do not let somebody come in and use that margin as a basis for luring our business customers away. That is not competition; it is the forced reallocation of market share.

As you know, GTE is not subject to the Telecommunications Act's section 271 restrictions on Bell operating companies' entry into long-distance markets. GTE can enter long-distance markets now. But the result of section 271 is, in my view, an outrage. The long-distance companies are cherry-picking away the high-value customers of the regional Bell operating companies, either through facilities or through unbundled network elements eventually, and are not creating competition for residential service. Meanwhile, the regulators have not dealt with universal service. So, as the value is being drained from the high-value customers of the regional Bell operating companies, those companies are being blocked from entering long-distance markets. Such an untenable situation outrageously distorts the act.

Let me just say a little bit about litigation. Some call GTE a world-class litigator; others claim that the firm scorches the earth. But GTE has undertaken no discovery, issued no interrogatories, taken not one deposition, made no requests to any state to stay the competitive process, and sought not one injunction. What GTE has done is to present serious legal questions about the ways in which the act is being interpreted. So far our firm has won. That does not mean that GTE is being unreasonable.

Before the Eighth Circuit decision, the FCC tried to preempt the process by mandating a certain set of prices and rules to the states. Before the Eighth Circuit had a chance to decide that case on the merits, the states had to act. The states entered interim decisions. In many cases they were distorted by or followed the FCC rules. Many of the arguments GTE was making in the Eighth Circuit that were still pending had been put into action at the state level. And, obviously, to preserve GTE's legal position—because what the Eighth Circuit would say about who had jurisdiction or about some of the issues of shared jurisdiction between the federal government and the state was unknown—the firm filed for judicial review of those rules in many states.

Since that time GTE has won just about everything it wanted to win in the Eighth Circuit. As a result, many of the issues raised in filings in the federal district courts are now moot. Many of the other issues raised in filings are moot because the long-distance companies have not been able to take advantage of the interim rules that were in effect. For example, nationwide, GTE received fifty interconnection orders from AT&T, all for customers who happened to be AT&T employees. On the basis of the projections of the long-distance companies, GTE set up centers to handle resale and unbundled network element business. Some 600 people are sitting there like the Maytag repairman because they are not getting any requests.

But GTE has used litigation and will continue to use litigation to get what it thinks are the fair and proper rules and what Congress intended under the Telecommunications Act. Judicial review of administrative action is the normal way that administrative policy is set. I cannot think of any significant federal rule that has not been reviewed in federal court.

But GTE is not relying on litigation to ensure that the rules and the act are not distorted. GTE is out in the marketplace doing what it thinks the act intended. GTE was the first large local exchange carrier to set up a competitive local exchange carrier, to carry the competition outside GTE's boundaries. GTE bought BBN, has been expanding the data business with an investment in the Qwest network, and has been pursuing video. Obviously, GTE has entered the long-distance market. That activity exemplifies the vision of the Telecommunications Act.

So GTE believes that the marketplace has solutions to the challenge of the Telecommunications Act, which ultimately was breaking down the product-line and geographical silos that existed. The act is creating one borderless telecommunications market so that companies can serve all markets and all customers with a full range of products.

For many companies, mergers may be the most efficient way of breaking down those silos and becoming full-service, national competitors. So it seems to me that the Telecommunications Act should lead to mergers and consolidations. For that reason, I believe, GTE's bid for MCI effectuated the Telecommunications Act and was very procompetitive, because the two companies are complementary, not competitive. On the other hand, GTE opposed the MCI-WorldCom deal on the grounds that it was anticompetitive—a classic horizontal merger between existing competitors. In GTE's view, an MCI-WorldCom merger would have severe anticompetitive impacts on both the long-distance and the data markets, and it would not in any way enhance local competition. Merging Brooks Fiber and MCI Metro—two competitive local exchange carriers operating in downtown areas and cherry-picking from the regional Bell operating companies—does not enhance local competition.

In sum, in the wake of the Telecommunications Act, nothing is antithetical about mergers and consolidation. Mergers are to be expected and, in appropriate cases, approved.

Delivered November 13, 1997.

8

Putting "People" in the Public Interest

F. Duane Ackerman

Is the Telecommunications Act of 1996 broken? My answer is, "No, but . . . ." The *but* portion addresses the fact that the act seems to be bending to the breaking point. The way the previous Federal Communications Commission—under the chairmanship of Reed Hundt—and other regulators have implemented the act has created two tracks: a fast one and a slow one. The fast track is for business customers to gain the benefits of competition; the slow track is for residential customers. Intended or not, the regulatory policy of the past twenty-two months amounts to an industrial policy that is not what Congress wanted.

WorldCom is emblematic of what is happening on the fast-track side of the equation. According to PaineWebber, with MCI, WorldCom would have greater revenues and freer cash flow than any regional Bell operating company. Measured by the reach of its networks, WorldCom will become the leader in technologies—namely, in data networks, which are for the most part the new world of communications. Today, companies are constructing new networks specifically designed for their business customers' data traffic, and it is easy to see why they do so.

Data revenues are expected to double over the next three years. They are growing at approximately five times the pace of revenue for traditional voice networks, which operate largely for consumers. MCI has said that its data traffic is already surpassing voice traffic. Analysts at BellSouth predict that within a decade voice—as measured in bits—will amount to only 10 percent of communications traffic. Meeting demand for data communications is the priority for major telecommunications companies. The opportunities and risks in meeting that demand are profound.

Clearly, many believe that WorldCom is well positioned in the data network game. It controls local fiber-optic rings in the business centers of the largest U.S. and European cities. It could thus become the leading provider of integrated services—including local service—to the U.S. business market, the leading competitive service provider in Europe, and the unrivaled leader of the Internet. There is little doubt why George Gilder welcomed the reign of "King Bernie" and crowned him as the successor to John D. Rockefeller.

My point is not to discuss the merits of the WorldCom-MCI deal, important as that is, or whether WorldCom will succeed. And we certainly do not want to say, "This deal should not happen," because it represents the way free markets function. My point is that regulators have put business customers on a fast track to competition while they have diverted residential consumers—the group regulators need to benefit—to a slow track.

Some policymakers at the White House, the FCC, and the Justice Department understand the potential of fast-track companies; they believe that those companies will drive innovation and economic growth. In short, the regulators want fast-track companies to be market-driven. Those companies are like WorldCom: they serve businesses—especially highly concentrated business areas—downtown, in office parks, and in areas along the interstate perimeters of our cities. They serve those businesses with data networks, intranets, extranets, and other Internet-based technologies that will probably include Internet telephony. Those fast-track companies do not mind selling plain old telephone service to businesses, either.

Other fast-track companies like Cisco Systems, Vocal Tech Communications, ITXC Corp., and Qwest are developing new technologies to challenge the paradigm of existing networks. Those companies are not often brought into the discussion about the Telecommunications Act, however. In addition, scores of other competitive local exchange carriers such as ACSI, Teleport, Brooks Fiber—now WorldCom—ICG, and Intermedia Communications are now reporting triple-digit revenue and access-line gains from their business customers.

That the business customer is frequently solicited and well served in no small measure results from the regulators' view that the massive, global change that fast-track companies initiate is good and should not be tampered with.

What, then, of consumers? What track are they on? How do we get residential consumers into the competitive circle? That is the fundamental question for regulators and policymakers working on the Telecommunications Act. In fact, that is something the new FCC chairman made reference to recently in Boston when he talked about how we cannot allow the information highway to bypass rural America and distressed inner cities.

While the fast-track companies are providing business customers with their sophisticated, data-network–based intranets, consumers are not a competitive focus of those same companies or of their networks. As a result, a very real risk is that today's voice networks—the same networks that overwhelmingly serve American consumers—will be marginalized and consigned to low-tech status as many companies exclusively chase the business telecommunications dollar and brush everyday consumers aside.

Disturbing data points have already surfaced. A November 1997 article in *USA Today* reported that AT&T's effort to "sell local phone service to consumers has all but stopped." That business decision is from a company that once vowed to "take 30 percent of the country's $100 billion local market." That is not good news for American consumers and the networks that serve them.

The regional Bell operating companies are not surprised by those reports because they always doubted the claims of those entrants into local telephone service markets who said that they would serve all customers, including residential. In fact, that is precisely the reason that Congress established a provision for the Bell operating companies to enter the long-distance markets if the long-distance companies ignored certain local markets.

The regional Bell operating companies have both the desire and the motivation to bring benefits not just to businesses but to individual consumers. Those benefits include lower long-distance prices and better services, as the regional operating companies evolve their voice-circuit networks to data networks for businesses and residences. Those are the same kind of data networks that the fast trackers are building only for their business customers.

The regional Bell operating companies must use their intellectual and investment capital to transform those networks. But doing so requires a huge investment that cannot be justified unless all markets are open to competition, including long-distance markets.

All consumers may not be fascinated with the evolution of voice networks to data networks, the rise of intranets and extranets, and their implications for communications policy and the economy. Consumers are, however, interested in communications at home—the telephone and ancillary features such as voice mail and caller ID. And more and more they are interested in advanced services, and many eagerly await high-speed access for the Internet.

Likewise, for many policymakers in Washington, residential service is what matters. There is no doubting the sincerity of the desire for local residential competition. The question is how to establish that competition. Some regulators want to manufacture competition by encouraging competitive local exchange carriers to make residential customers their priority. But other regulators and policymakers are beginning to recognize that subsidized pricing—which is deeply embedded in our social fabric—

makes the local residential market highly unattractive to many would-be competitors.

Those disincentives—powerful on their own—are coupled with a somewhat perverse incentive that further affects the behavior of long-distance companies. The regulatory policy that created the perverse incentive rewards long-distance companies for staying out of local, residential service because staying out of that market makes the entry of local exchange carriers into long-distance markets more difficult to justify.

To see how the perverse incentives work, consider the example of a company in Atlanta that can build a 100-mile network from Atlanta's midtown up through Buckhead and on toward the north perimeter. That company can sell local, local toll, long-distance, Internet access, and other services to businesses—a sales volume that represents about 30 percent of the total business communications spending in the entire state of Georgia. WorldCom, through MFS, has over 150 miles of fiber in that area. MCI has built more than eighty fiber miles. ICG, based in Denver, is building a 100-mile network. Even though large residential apartment buildings exist right alongside their networks, those companies do not market services to residential consumers.

Building only where the business dollars are is happening not only in Atlanta but in Miami, Charlotte, Memphis, Orlando, New Orleans, and every other major business center in BellSouth's nine-state region. Businesses in cities such as Pensacola, Mobile, and Chattanooga can put all their business communications traffic on a competitive network. In fact, competitors now operate more than 100 facility-based networks across BellSouth's region—from Winston-Salem, North Carolina, to Lake Charles, Louisiana, and from Louisville, Kentucky, to the Florida Keys. Those networks offer businesses competitive choices in almost every city served by BellSouth.

Why, then, should those companies put their long-distance revenue at risk by marketing local residential service? Under current policy, they will not do so because mass-market local entry by those companies will allow BellSouth to offer long-distance service, either immediately or at least on an expedited basis. Indeed, one can argue that under current policy the long-distance companies have a compelling incentive not to enter local residential service markets in any significant way as long as not entering those markets prevents the regional Bell operating companies from entering long-distance markets. In fact, the long-distance carriers do not offer residential service in any significant way.

Conversely, the way to create incentives that would prompt long-distance companies to offer local residential service and to switch consumers from the slow track to the fast track is to allow the local companies to provide long-distance service, just as Sprint, GTE, Southern New England Telephone, and some 1,500 other local companies have been allowed to do. When the regional Bell operating companies can offer long-distance service, their competitors can view local customers not as mere $20-a-month consumers, but as potential $200-a-month customers. On a monthly average, consumers spend $20 for local service, $35 for long-distance service, $20 for Internet access, and $50 for wireless service. With the addition of security alarm service and entertainment, a customer who spends $20 a month may in a competitive market spend $200 a month. That provides the incentive to enter local residential service.

Can that work to the consumers' benefit? Since Southern New England Telephone entered the interstate market in 1994, it has reduced residential long-distance rates by nearly 18 percent. In response, AT&T has petitioned for authority to reduce its long-distance rates specifically for Connecticut. AT&T is actually fighting for residential customers; and consumers are experiencing the benefits of competition.

Along with long-distance rate reductions, what comes with competition? Perhaps companies will package integrated services like local, long-distance, Internet, and wireless to retain those customers. Perhaps the long-distance companies will make available to consumers some of the sophistication they are offering their business customers. Perhaps, as a result of that competition, consumers will be switched onto the fast track.

We can debate the outcome for another two or three years. In fact, some officials have argued outright: "Go slowly. These things take time." Regulators said the same thing in the early 1970s with the advent of cellular technology. They took their time, allowed regulation to lead, and kept wireless technology out of the reach of average Americans for nearly a full decade.

The American people need not go through that experience again. Now, not later, is the time to put people back in the public interest. Why should consumers be patient when the long-distance companies are not being patient about taking business customer revenue for local, toll, and data services? Why should consumers accept the explanation that "these things take time" when it has not taken time for long-distance companies to solicit business customers with local service data networks? And why should consumers be tolerant when long-distance companies are being rewarded for staying out of local residential service? Why would anyone who knows the history of cellular telephony be patient?

The answer is that consumers and the regional Bell operating companies should not have to be patient and, in some areas, will not have to be. The South Carolina Public Service Commission unanimously approved BellSouth's application to enter the long-distance market. It also analyzed BellSouth's operating support systems and ruled that they will do the job for competitors.

Nothing in the Telecommunications Act mentions operating support systems. Instead, the act employs a competitive checklist. The reason for that list and its provisions, of course, was to create some objective standard by which to judge the regional Bell operating companies' progress toward helping others to enter the local-service market. Congress added that neither the FCC nor anyone else could expand the checklist.

The provision in the checklist that attracts so much attention is the one that requires the regional Bell operating companies to give entrants access to databases. BellSouth accommodated that requirement and found it reasonable that access to those databases be electronic so as to be nondiscriminatory. BellSouth spent the money and the time to accommodate all parties, and the South Carolina Public Service Commission agreed that the systems could do what they should do. BellSouth demonstrated the system to anyone who wanted to examine it. After the company's effort, time, and expense, the previous FCC—under Reed Hundt's chairmanship—issued the *Ameritech* decision, which subjected the regional Bell operating companies to more regulation than existed before passage of the Telecommunications Act.

Fortunately, the regional Bell operating companies are beginning to hear some encouraging things from the new FCC. That old regulatory phrase from long ago—"common sense"—is reappearing and is most fortuitous. Creating competition in the telecommunications market needs a common-sense approach. The current FCC now has an opportunity to bring the two telecommunications tracks together in South Carolina.

BellSouth has promised to cut long-distance rates in South Carolina and to offer the packages that customers want. If it is irrevocably clear that long-distance competition has begun, the long-distance companies should offer packages to all customers as well. As the South Carolina commission has asserted, BellSouth's entry into the long-distance market is the only way to bring the long-distance companies into the local residential market. Customers—business and residential—would be the winners. That is a measure of the public interest that should get more weight.

Some regulators believe that it will be years before a viable, alternative technology emerges to compete against traditional landline service. Regulators also clearly think that it will take considerable time to overcome the problems that flow from subsidized pricing. Contentious issues include how universal service will be funded and how it will be distributed. Not everyone will be happy with any resolution of those issues, but a resolution would establish a base. To delay and to tell residential consumers to be patient is not an alternative. It is a sure way to keep the benefits of competition for consumers on the slow track.

If regulators really believe that their first responsibility is to consumers, they cannot allow the current policy of a fast track for business and a slow track for consumers. We must change that industrial policy on behalf of consumers so that they too can enjoy the benefits of the fast track: market forces must rule. And South Carolina is a wonderful place to begin by making right the disparity consumers are being asked to endure today—a disparity in savings, choice, and access to the fast competitive track.

Delivered November 20, 1997.

9

Foxes, Hedgehogs, and Federalism: States Implement the Telecommunications Act

Bob Rowe

Each contributor to this valuable series of assessments of the Telecommunications Act of 1996 has added important brush strokes to what the American Enterprise Institute's Studies in Telecommunications Deregulation will be able to offer as a portrait-in-progress of local competition. Whether the portrait will be a Rembrandt composition or an M. C. Escher maze remains to be determined in the emerging telecommunications markets.

It is appropriate to include a view from the states. Almost everyone agrees the public policy battle for customer choice, innovation, and investment will ultimately be won or lost in significant measure in the states and especially at state public utility commissions. Indeed, state commissions, given the tools, are committed to seeing the battle won for the good of their citizens, their markets, and their states.

From Archilochus to Madison

The title of my contribution honors Sir Isaiah Berlin, whose famous essay quoted the Greek poet Archilochus: "The fox knows many things, but the hedgehog knows one big thing." The fox pursues several ideas at the same time, even when they may be inconsistent. The hedgehog relates everything to a single idea and pursues it persistently. Aristotle was a fox. Plato was a hedgehog. In telecommunications policy we must recognize the value of diverse perspectives and respect a little bit of uncertainty.

The Telecommunications Act was a big hedgehog of an idea. It had to be. We were sixty years adjusting to the old regime, tectonic pressures were building, and Richterlike action was required. The act set high standards for the Federal Communications Commission and for state commissions. To its credit, the FCC largely did what Congress instructed and did it on time. During the first phase of implementation, the FCC was the perhaps necessary hedgehog in the federalist system, resolutely focused on the one big idea of implementing a national competition policy. The states have some quills to show for it.

The act, however, is profoundly federalist in its structure. It speaks directly to states and state utility commissions. Congress explicitly affirmed state commission jurisdiction over intrastate rates and services; set out a continuing role for federal-state joint boards; affirmed state commission responsibility for customer service; required consultation with states on Bell operating company entry into the in-region long-distance market; described the role of state universal service funds; required cooperation on audits; rejected (with the support of the National Association of Regulatory Utility Commissioners) state and local barriers to competition; and—crucially—gave state commissions the lead role in arbitrations, mediations, and contract approval, including implementing pricing and interconnection policies consistent with congressional intent. I say crucially because, if we are successful, those agreements and orders will shape the new structure of telecommunications markets.

State commissions are the foxes in the federalist structure. Nothing is more frustrating to a hedgehog than contending with a fox: worse is contending with more than fifty foxes, each with its own economist, engineer, customer service specialist, and lawyer. Pity the hedgehog.

Competition—Are We There Yet?

We generally agree that we are moving from stand-alone parallel networks (switched wire, wireless, cable, and private line, with the switched telephone network as paramount), through a linchpin network based on the existing switched wire network, toward—we hope—a network of networks, all of which are interconnected with one another and within which—again we hope—none has strategic advantage over another. Correct public policy requires knowing where we are and—without locking in technology choices that should be made in the market—having some idea of where we think we are going.

This AEI series has done a marvelous job helping policymakers understand where we are. Contributors have implicitly employed two dominant economic lenses through which to view telecommunications markets. The names are not precise, but let us call them the idealist, laissez-faire and the strategic, industrial organization lenses. I apologize to economists of both persuasions.

We have much to learn from longitudinal studies of telecommunications markets, for example, from those tracking the consolidation of U.S. local providers in the early 1900s, followed by more recent moves to reopen various telecommunications markets.[1] Our rallying cry could be "Remember the Hush-A-Phone." We can also learn from comparative studies of telecommunications policies and markets in other developed nations including Great Britain, New Zealand, Canada, and the countries in northern Europe.[2] It is striking how one can describe the same market differently depending on the economic lens through which he views it.

The first lens, the idealist, laissez-faire lens, focuses on the dynamic actors and technology already in the telecommunications market, the ever-lowering barriers to entry, a high level of existing competition in some specific areas, and the real threat of potential competition as further discipline on current market participants.[3] Peter Huber described the hedgehog–eye view through that lens when he called for a complete return to common law.[4]

The second lens, the strategic, industrial organization lens, focuses more on the strategic decisions of actors within specific market structures. It is associated with the analysis of structure, conduct, and performance and with measures of market concentration such as the Herfindahl-Hirschman index.[5] Assistant Attorney General Joel Klein's views in this series partly reflected that perspective, consistent with his antitrust responsibilities. It is impossible to pigeonhole anyone's views, as demonstrated by Huber's strong belief in antitrust enforcement.

Here is my attempt at an aphorism: "One who views the world through one lens only has no perspective and will likely bump his nose on the very object he is trying to avoid." That is another lesson inferred from Isaiah Berlin, although he would have said it much better.

We need to know the simple, raw number of customers who have picked a new local carrier, and we need to know how hard incumbents are working to develop efficient operations support systems, as Bell Atlantic's Ray Smith so well described. That is the first lens. We also want to know the percentage of each class of customers that has switched, as well as the market share for the major firms in each market. That is the second lens.

I suggest some tentative lessons from all those studies. First, proactive, long-term policies are often required. Second, competition-enhancing policies may be both large, enacting the Telecommunications Act, and small, deciding one of the many specific issues that arise in a state proceeding. Third, firms cannot realistically be expected to relinquish voluntarily a particular advantage without gaining something in return. Recall how successfully local carriers have been negotiating end-to-end interconnection agreements when it has clearly been a "win-win" situation for them and their customers. Fourth, with that in mind, removing bottlenecks is necessary but not necessarily sufficient. Fifth, to know what is happening, we need both qualitative and quantitative information. That is something that state commissions, which are close to customers and to markets, are uniquely able to develop. Sixth, the move from monopoly to competition will proceed through stages of tight market structure absent some significant technological change. Seventh, the public policy role is to establish flexible, market-appropriate policy, not to preordain specific outcomes. Finally, customer dissatisfaction, whether with slamming, service quality, or price, could derail the train. We need to pay attention to customers both because it is necessary and because it is right.

I also suggest some tentative lessons from experience in other countries. First, absent Schumpeterian innovation, market structure does not change so quickly as some might like. Second, market share moves more quickly for higher-volume than for lower-volume customers and more quickly for business than for residential customers. British Telecommunications, for example, has 72 percent of business customers and 88 percent of residential customers. Third, local loop companies do well competing against national firms, as we have seen in Finland: the loop is still a crucial asset. Fourth, prices will fall with competition and will fall more with more competition. Finally, on the basis of New Zealand's four-year trip to the Privy Council in London, lawyers, like the poor, will always be with us.

State Actions Implementing the Act

As were many state commissioners, I was dismayed by much of the general-interest press coverage of the Eighth Circuit's July 18, 1997, decision in *Iowa* v. *FCC*. The court invalidated the FCC's pricing rules on the basis not of an economic theory or a constitutional "takings" argument as advanced by the incumbent providers, but of state jurisdiction over intrastate rates and service under section 152(b).[6]

Reports of the death of local competition were greatly exaggerated. While the *New York Times* did not run the story on the obituary page, much of the general-interest press "got spun." To understand the true condition of local competition, it is necessary to look at what is happening at state commissions and in real, local markets.

Remember that the FCC pricing rules were stayed nearly a year earlier, in September 1996. During that time, with no binding national rules in place, state commissions went ahead doing what Congress told them to do. The National Association of Regulatory Utility Commissioners studied state-by-state information on the Telecommunications Act's implementation, including interconnection agreements, arbitrations, pricing proceedings, section 271 dockets, and all aspects of universal service.[7] NARUC found that states, without being required to do so by the FCC, had approved more than 1,100 new competitors, had approved more than 500 interconnection agreements, and had completed at least 180 arbitrations (with 115 more pending). In addition, the states had opened section 271 proceedings on Bell operating company in-region, long-distance entry in at least thirty-four states, had or have been defending more than seventy court challenges to their procompetitive policies, and had developed state universal service programs. Finally, NARUC found that the states had opened proceedings concerning operations support systems and had developed competition-oriented customer information and consumer protection programs.

An exclusive focus on national implementation, national markets, or national firms tells one important part of the story but misses many equally important parts. Such a focus misses the courageous procompetitive work being done by the Iowa Utilities Board and the innovative efforts of the Colorado Public Utilities Commission, to name only two of many. That focus misses the interesting entry strategy of midsized Iowa-based McLeod USA. It misses the story Representative Bliley noted when he spoke of the very small but growing Mid-Rivers Telephone. Mid-Rivers has captured significant telecommunications market share from U S

WEST and cable market share from TCI in Terry, Montana, by providing customers what they want when they want it, with U S WEST and TCI responding in kind.

Through the idealist, laissez-faire lens, the numbers I cited are great news! Through the strategic, industrial organization lens, the response is, "Not so fast! Important questions remain unanswered."

The FCC and the states have managed to work together cooperatively on many of the items on my list, with disagreements clearly identified and confined.

If the Eighth Circuit's July 1997 decision was not a death blow to competition, and if the states and the FCC are already working well together on many fronts, why is competition not developing more quickly? At least five reasons are apparent. First, new competitors blame the current providers for delay and obstruction. Indeed, both the states and the FCC are dealing with critical enforcement issues. In fairness, it is important to note that "delay and obstruction" can be a two-way street. Second, current providers say that new competitors are not serious about competing for local business, or that they only want to "skim the cream" of lucrative large business accounts, or that they just do not understand how complicated local service can be. New entrants respond that they need prices and other terms that allow them to make their business case. Third, local competition is technically more complex than long-distance competition and requires staggering capital investments. Fourth, widespread provision of local phone service is enormously capital-intensive. Fifth, absent introduction of a market-changing technology, market share usually changes slowly.

Some consequences of public policy decisions are not foreseeable. For example, cable and phone companies have not yet entered one another's markets to the extent anticipated. On the other hand, when the Telecommunications Act passed, few people outside the industry laboratories were thinking much about fixed wireless service. Currently, there is lively debate about whether mergers and acquisitions were a foreseeable consequence of the act or even of the FCC's price-cap and access-reform orders.

The Positive State Agenda—More than Jurisdiction

In the Telecommunications Act, Congress spoke directly to the states. State commissions take that direction seriously and value the sustained congressional interest in what the states are doing to open local markets, promote competition, and benefit customers.

As Congress and the FCC have recognized, state commissions introduced many innovative policies to pioneer local competition: unbundling and interconnection, various forward-looking pricing approaches, retail-wholesale separations as a way to align structure and desired conduct, and even wiring schools and libraries. Some work well; others need more work; some perhaps just will not work. That is the diversity benefit of federalism, something foxes may be better able to appreciate than hedgehogs.

State commissions' concern to protect their authority over intrastate rates and services is not "jurisdiction for jurisdiction's sake." Rather, the concern is to preserve the commissions' ability to carry out positive, proconsumer, and procompetition policies appropriate to their markets.

All that interesting work comes together in NARUC and its Communications Committee and Staff Subcommittee. In the years before the act's passage, the Communications Committee undertook a comprehensive analysis of how best to implement local competition, which is still a valuable source book.[8] That report informed NARUC's legislative positions as the act was developed—positions that were thorough and were thoroughly procompetitive. At the same time, NARUC supported reaffirmation of state jurisdiction over intrastate rates and service and also supported prohibition of barriers to entry. NARUC's position against barriers to entry was a watershed.

Three times a year, the Communications Committee develops and passes resolutions that it forwards to the

NARUC Executive Committee for final action. Those resolutions guide NARUC's work with Congress, in the courts, with the FCC, and on other substantive matters. For example, in November 1997, NARUC passed resolutions concerning slamming, Internet taxation, reciprocal compensation of Internet service providers, court jurisdiction over Telecommunications Act matters, operations support systems, jurisdictional separations, and the universal service high-cost fund. Each year, the committee develops a plan to guide the work of its very active commissioner and its staff policy subgroups.

Universal Service High-Cost Fund. This is a subject that divides lower-average-cost and higher-average-cost states. States tend to see themselves as either burdened or benefited by the federal high-cost program. But states have been working together, with more success than most observers expected, to recognize one another's legitimate concerns and devise approaches consistent with the Telecommunications Act.[9]

Section 271 Bell Company In-Region Long-Distance Service. States went back to the Eighth Circuit on the narrow issue of whether the FCC could do indirectly under section 271 what the court said it could not do directly: impose a specific pricing methodology. The Eighth Circuit issued the writ of mandate, the first step, but has not yet reached the substance of the states' claim. The states' disagreement focuses exclusively on the pricing section of the FCC's *Ameritech* order. States view their action as seeking enforcement of the court's prior order and nothing more. As Senator McCain pointed out, states are already largely using forward-looking methodologies. More recently, the Department of Justice has suggested that a variety of appropriate forward-looking methods exist.

Apart from pricing, section 271 provides a good example of cooperation among state commissions, the FCC, and the Department of Justice: the three collaborated to develop a statement on best practices for processing section 271 cases. Both the Justice Department and the FCC have systems of regional contacts with state commissions, something NARUC encouraged. I hope we can strengthen that coordination. In the near future NARUC hopes to provide state commissions with a template listing the key items identified by both the FCC and the Justice Department that states may use as they consider appropriate in processing those cases and developing high-quality records. The point is to support state efforts to do excellent work.

Consumer Issues. Complaints are increasing about slamming, billing for services never provided, use of confidential customer information, and other abusive practices. In some regions telecommunications customer service and network service quality have also become major concerns. Several states have adopted competition-oriented telecommunications customers' statements of rights. The Communications Committee has long made consumer issues a priority, and the National Regulatory Research Institute provides important research and assistance in that work.[10] In 1997 NARUC established an ad hoc Committee on Consumer Affairs that is actively addressing consumer issues in all utility industries.

Mergers and Acquisitions. Mergers can be either pro- or anticompetitive. The questions get especially complicated where markets are being opened for the first time.[11] Antitrust analysis can be quite useful, but traditional antitrust procedure may not be so efficient or effective as needed. NARUC and the National Regulatory Research Institute have been examining the application of antitrust analysis to regulatory work for several years.[12] Another area where Congress has evinced a strong interest, that analysis invites state-federal cooperation.

The Communications Committee and state commissions generally are increasingly consumer-focused, serious about understanding markets, and supportive of competition. State commissions look less and less like the Interstate Commerce Commission and ever more like the Federal Trade Commission.

With all that good work and goodwill, why do federal and state telecommunications regulators not get along? I have saved that topic for last.

State-federal tension over telecommunications policy is as old as pegboard switches. It has generated a long line of case law and law review articles.[13] A healthy federal-state tension was intended by the framers of the United States Constitution and carried over into both the 1934 Communications Act and the 1996 Telecommunications Act. Still, over the past several years, we have done much good work together despite our jurisdictional and substantive disagreements. With many of the act's deadlines now behind us, it may be possible to concentrate some attention on improving how we work together. FCC Chairman Bill Kennard made clear that that was his desire when in his first speech after confirmation he invited states to enter a "Magna Carta" and wisely left the drafting for a joint exercise.

Isaiah Berlin argued for the ability of an individual to prevail against the tides of historical determinism—his "great-man theory." Perhaps this is a time when individuals of goodwill can, consistently with the act as interpreted by the courts, rechannel that rivulet of history. We just need good people, not great men. We must all think like foxes rather than hedgehogs to do that.

I suggest three elements: first, a general statement of an agreed-upon approach; second, a list of desirable least-prescriptive practices applicable to common issues; and third, a rolling list of the substantive areas where we must work together.

Suggested General Approach. State commissions and the FCC possess complementary strengths. They should work together to take full advantage of each party's attributes.

State commissions are close to the customers and have strong consumer affairs programs. They are close to local markets and have developed quantitative and heuristic methods ("listening to customers") for evaluating the structure of those markets. They have fact-based proceedings and are able to analyze and act on complex records. They also benefit from experience with other industry restructurings, including natural gas and electricity.

The Federal Communications Commission has a national and even global perspective. The agency is concerned with all communications media—wireline, wireless, cable, and satellite. The FCC has a strong policy emphasis. It works directly with the companies, collects industrywide data, and possesses excellent analytical abilities.

Consistent with the federalist scheme Congress established in the Telecommunications Act, FCC actions affecting states should be undertaken in the most flexible, least prescriptive way possible. In areas where national standards are desirable and are consistent with the act, they should be implemented in ways that do not preclude state experimentation or state efforts to exceed those standards. Generally, the most desirable approach is to strengthen state procedures.

Suggested Practices. Methods to accomplish the general approach include developing procedural "best practices" guidelines, convening federal-state joint boards, and developing substantive models or standards that the states may consider in formulating state-specific policies. In addition, the FCC and state commissions should develop data and analysis. The FCC should participate directly in key state proceedings and intervene in judicial proceedings challenging procompetitive state commission decisions. State commissions should participate in proceedings challenging procompetitive FCC decisions that do not raise jurisdictional issues.[14] Finally, the FCC should engage in judicious and limited preemption of state and local actions that constitute "barriers to entry" under section 253, after full exhaustion of state-level remedies that may be available. Something like an "exhaustion of remedies" approach would strengthen state-level proceedings and promote administrative efficiency.

Congress set the bar high for the FCC, for state commissions, and ultimately for competitors in the marketplace. It, along with the Clinton administration, has shown exceptionally informed and thoughtful interest in how we carry out our duties. We have not yet accomplished the job we were given, but we are now seeing the early fruits of our efforts.

Rather than pull mutually inflicted hedgehog quills from all our hides—public and private alike—perhaps we can use just one quill to initial the Magna Carta Chairman Kennard proposed.

Delivered December 1, 1997.

Notes

Unless explicitly identified as such, the views expressed are not those of NARUC or the Communications Committee and do not represent the views of the Montana Public Service Commission. They are not intended as comment on any pending or future proceeding before the Montana commission.

1. NARUC's research arm, the National Regulatory Research Institute, for example, has published Kenneth Costello and Robert Graniere, *Deregulation-Restructuring: Evidence for Individual Industries* (Columbus, Ohio: National Regulatory Research Institute, 1997). The institute is a valuable source of analysis and information on telecommunications and other utility issues.

2. Good comparative discussions may be found in William H. Melody, ed., *Telecommunications Reform: Principles, Policies, and Practices* (Technical University of Denmark, 1977), and in Robin Mansell, *The New Telecommunications: A Political Economy of Network Evolution* (London: Sage, 1993). See also the special telecommunications survey in the *Economist* (September 13, 1997) and in David Gabel and William Pollard, *Privatization, Deregulation, and Competition: Learning from the Cases of Telecommunications in New Zealand and the United Kingdom* (Columbus, Ohio: National Regulatory Research Institute, 1995).

3. Addressing the timely issues of federal and state policy, Kenneth Gordon and Thomas J. Dustenberg provide an excellent example of that approach in *Competition and Deregulation in Telecommunications: The Case for a New Paradigm* (Indianapolis, Ind.: Hudson Institute, 1997).

4. *Law and Disorder in Cyberspace: Abolish the FCC and Let Common Law Rule the Telecosm* (New York: Oxford University Press, 1997).

5. Good examples are Melody, ed., *Telecommunications Reform,* and Mansell, *The New Telecommunications.*

6. A good example of the kind of coverage required is Albert R. Karr, "State Utility Regulators Bid Goodbye to Coziness with Industry and Enter the Competition Wars," *Wall Street Journal,* December 26, 1996, p. 32, concerning the Georgia Public Service Commission. Of course, that is the kind of detailed coverage much of the specialized telecommunications press routinely provides.

7. *Telecommunications Competition 1997* (Washington, D.C.: NARUC, 1997).

8. *Local Competition Summary Report* (February 1996).

9. Two reports prepared for the Communications Committee are worth noting. The Telecommunications Industry Analysis Project prepared *Options for the Universal Service Fund* (October 15, 1997), which models several alternatives for contributions into the high-cost fund. That project provides valuable, neutral analysis on a range of universal service and other telecommunications issues. The Communications Committee also established an ad hoc work group on high-cost fund issues that produced *High-Cost Support: An Alternative Distribution Proposal* (November 3, 1997), which focuses on disbursement from the fund as a way to provide reasonable comparability of urban and rural rates and service while controlling the total size of the fund.

10. See Vivian Watkins Davis et al., *Telecommunications Service Quality* (Columbus, Ohio: National Regulatory Research Institute, 1996); Raymond W. Lawton, *Survey and Analysis of the Telecommunications Quality-of-Service Preferences and Experiences of the Customers of Ohio Local Telephone Companies* (Columbus, Ohio: National Regulatory Research Institute, 1996).

11. William Landis and Richard Posner noted the difficulty of directly applying market measurements to rate-regulated markets in "Market Power in Antitrust Cases," *Harvard Law Review* 94 (1981): 937, 975–76.

12. Edwin A. Rosenberg, *Telecommunications Mergers and Acquisitions: Key Policy Issues and Options for State Regulators* (Columbus, Ohio: National Regulatory Research Institute, 1997). An earlier publication in the series is David Chessler, *Determining When Competition Is "Workable": A Handbook for State Commissions Making Assessments Required by the Telecommunications Act of 1996* (Columbus, Ohio: National Regulatory Research Institute, 1996).

13. See, for example, Jonathan Jacob Nadler, "Give Peace a Chance: FCC-State Relations after California III," *Federal Communications Law Journal* 47 (April 1995): 457.

14. NARUC has a fiduciary obligation to its member states that it must consider when it acts.

10

Out of the Courts and into the Market: Wouldn't It Be Great?

John D. Zeglis

Why is it that almost two years after the Telecommunications Act became law, the benefits of choice have barely touched America's local phone customers—and then only in the most ephemeral form? That is the question I want to address.

First of all, I contend that the act is not broken and does not need to be fixed. It is, however, under siege. Led by GTE and the Bell operating companies, the local exchange monopolies appear to be running a war of attrition against the very legislation they supported in 1996. From legal suits to regulatory challenges to inexplicable delays, the Telecommunications Act is not being implemented as Congress intended.

As that war of attrition is pressed forward, two fictions are being spread: that the act is flawed and that AT&T does not really want to enter the local service business. I shall set the record straight on both those counts by first considering the Telecommunications Act itself.

The framework of the act is as near to correct as the legislative process allows. It gets the incentives right for at least 80 percent of the industry. First, the Bell companies open their local monopolies; then—and only then—may they enter the long-distance market. Neither does the act leave the methods of opening the local exchange markets to question: It gives a detailed checklist of sensible operating conditions that have to be met. By doing so, the act requires incumbent monopolies to make their networks available to new competitors.

The act set up three modes of entry, most likely to occur in sequence, for the massive job of introducing competition to the century-old monopoly in the local phone service market. First, new competitors can enter that market through what the industry calls total service resale. A new competitor buys the local monopoly's complete service package—with no substitutions, no customizing, and no real economic potential—because the new competitor gets only a token discount from the retail price and still has to pay whopping access charges to the incumbent local company for any long-distance service connections to the customer.

The fundamental advantage of total service resale is that it should enable a new competitor to enter the business rapidly, albeit on a limited basis. That method of entry was to be quick and dirty, to jump-start competition where there had not been any for almost a century. Total service resale offered a chance for easy market entry, followed by more meaningful forms of competition.

We were only half right, however. Total service resale has not proven to be quick, but very, very dirty. It has been difficult to get service sold and delivered from incumbent local companies to new competitors. As a result, the fundamental advantage of total service resale has been lost.

Thus, the best hope for seeing real competition in the near future lies with the second option: unbundled network elements. Those elements are the parts of a local network—from the line that runs to the customers, to capacity on the local company's switch, to operator services and other forms of support. A new competitor that buys unbundled network elements can lease as many or as few elements of the local network as needed to get started. Or the competitor can add its own elements over time, as needed. If new entrants lease all the elements and then use them for local service, we call that method of entry the unbundled network elements platform.

The important point is that a new competitor is supposed to be able to buy unbundled network elements at prices based on cost, including a fair return, without thereafter paying grossly inflated access charges to the incumbent local company for any long-distance service that runs over the local company's network. Unbundled network elements are to be sold to competitors on the basis of the forward-looking costs of an efficient firm. So they essentially replicate a competitive wholesale market where all retail competitors start off on an equal-cost footing—even when one of the retailers in the market is the very incumbent local company that is acting as wholesaler to all the competitors. Competitors win on the basis of what they can do for customers, not on the basis of a locked-in cost advantage.

Although unbundled network elements are the only practical route for broadly delivering benefits of competition to residential and most business customers within the next few years, over time AT&T and other long-distance carriers will undoubtedly build some local network facilities of their own.

That brings me to the third option cited in the act: facilities-based competition. Under that option a new entrant can wholly or partly replace some—but not necessarily all—of the unbundled elements it leases. That option is not a quick route to local competition. Matching the embedded network capacity of local companies would involve a massive investment of money and would take years to do. Moreover, for technology to create realistic alternatives to replace elements such as residential loops will take a long time. AT&T cannot afford to wait; neither do its customers want to wait—considerations that bring me to the fiction about AT&T's willingness to provide local service.

AT&T's Commitment to Compete in the Local Service Market

Local monopolists are spending heavily to promote the fiction that AT&T does not really want to compete in providing local service. Simply put, that is ridiculous. AT&T badly wants to enter that market. In bottom-line terms, the provision of local service is a strategic and economic imperative for AT&T. Strategically, offering local service is a crucial part of the integrated service packages on which we are counting to create customer value in the future. AT&T wants to be the industry leader in providing services over any distance, in any form. Local service is absolutely critical to delivering on that strategy.

With respect to financial incentive, the local service market is a $100 billion-a-year market virtually untouched by serious competition. AT&T's long-distance customers alone would be a vast new revenue stream if AT&T could sign them all up for local service as well.

AT&T's commitment to provide local service is total, even if local exchange companies run commercials questioning our commitment. AT&T filed for permission to provide local service in all fifty states the same month the Telecommunications Act was signed and has invested billions of dollars in local market entry. AT&T will invest more in local service in 1998, even at a time of major cost reduction and sharply prioritized spending—so long as entry conditions allow AT&T a fair shot at earning a return on the investment.

Psychologically and emotionally AT&T made the big leap in the early 1990s after deciding that the firm could improve the status quo. The firm could find a way to give customers choice where none had existed and a way to bring down government-imposed barriers: AT&T would enter local service markets, and the Bell operating companies would enter long-distance markets. AT&T went from just saying no to asking how the firm could make those changes.

The answer came in the form of the Telecommunications Act of 1996. Having made that pivot and having staked a big part of the firm's future on it, AT&T is hardly feigning interest in the local market—quite the contrary.

AT&T is disappointed that its progress in entering the local market has not been quicker and wishes that it had more to show for a very large investment. AT&T has 400,000 customers through total service resale in eight states but still has not made much of a dent in the incumbent local carriers' market share. AT&T has had to scale back marketing because of the difficulties encountered in getting local exchange companies to handle the flow of orders the firm can generate. But the importance of local service and AT&T's commitment to it remain.

The problem is not in AT&T's lack of commitment to provide local service. The problem is with the companies most responsible for implementing the act. The firms running those advertisements know very well why competition has been slow coming to the local market. So far, they have managed to keep competition in the courtroom and out of the marketplace. AT&T must reverse that situation, which reminds me of Will Rogers's comment, "We are always saying, let the law take its course, but what we mean is, let the law take *our* course."

The Bell operating companies are following a strategy I could have been proud of as a Bell System lawyer. Because their markets have to be opened and because they are the ones who have to open their markets, they concentrate on legal tactics that get things into court where they can be stayed, delayed, or otherwise strung out and then blame the delays on the firms that want to enter local markets. The Bell operating companies frustrate public policymakers to the point where they throw up their hands and simply let local exchange companies enter the long-distance market.

Such a strategy will not succeed. U.S. public policymakers are made of sterner stuff. The Telecommunications Act is a good one, and the policy it supports is so magnificently proconsumer that it can have no other outcome. AT&T *will* get those markets open. The incumbent local carriers will test the will of public policymakers. But the policymakers will respond by insisting that those carriers deliver what the public wants: choice.

All of which is to not to say that there is not a bit of *déjà vu* here. Recall the movie *Back to the Future* where Michael J. Fox and Christopher Lloyd went time traveling in a souped-up DeLorean, changing history to suit their own needs. I sometimes worry that we are on a similar backward-looking journey and that our DeLorean has two possible stops. One is back in the 1970s, in the chaos and false starts that preceded competition in long-distance service and telecommunications equipment. The second stop—if the first one succeeds—could be all the way back to the 1950s, with a series of monopolies reminiscent of the old Bell System.

Lessons of Long-Distance and Equipment Competition

To understand what I mean, let us travel back into the late 1960s. Competition came knocking on the door of the Bell System's long-distance and equipment markets. Predictably, the Bell System tried to shut the door tight. As counsel for the Bell System, I was in the thick of the battle. The Bell System did everything it could to fight competition on multiple fronts. A favorite tactic was promoting piecemeal decisionmaking—by subject and by jurisdiction.

For many wasteful years, decisions were made by different jurisdictions, one segment of the market at a time, one facet of competition at a time. For years the Bell System was subject to no national, unified policy. It took MCI fifteen years to go from its certification in 1969 to its first dial-1 easy-access call in 1984 because policy and rules were made in too many courthouses and hearing rooms. That regime was a great time for lawyers, but it did not do much for long-distance customers.

It took the federal government to create and implement a new national policy. Judge Harold Greene acted under federal law to enforce the antitrust decree that required AT&T to divest its local telephone companies, and the Bell operating companies were obliged to treat all long-distance competitors equally. The FCC emerged as the national rulemaking body on equipment and long-distance competition.

With the benefit of hindsight, all parties can all be grateful. We now have a competitive long-distance market with more than 500 players, large and small—a market where competition has given customers price reductions of better than 50 percent since the Bell System breakup in 1984. That market spurred AT&T to advance the digitization of its network by nearly a decade—not because of a regulatory rule, but because the firm heard a "pin drop." Consider the 40,000 route miles of fiber optics in the AT&T network and the advanced technology of our competitors. The long-distance market today is proof positive of the adage that competition is the lubricant of innovation. But none of that would have happened in the long-distance market if the piecemeal efforts to introduce and nurture competition had not finally been brought together.

The Parallel with the Local Market Situation

When it came to introducing competition into the local phone service market, the Telecommunications Act learned from the 1970s and 1980s and gave America a huge advantage. The act gave the industry one comprehensive national policy—one set of rules for making competition happen. Where long-distance rulemaking converged only after many years, the Telecommunications Act gave that advantage from the outset for local service.

But we must look out for that time-traveling DeLorean loaded with lawyers from the local exchange companies—recklessly resolved to turn the clock back to the times in which public policy was made in jurisdiction A on Monday and reversed in jurisdiction B on Tuesday.

Eighth Circuit Implications

I am not creating science fiction or engaging in speculation. The *Back to the Future* strategy is written all over the Eighth Circuit; just read the decision. The Eighth Circuit has recreated the divided jurisdiction of the past by holding that the FCC cannot adopt national rules implementing important parts of the federal law. Instead, the Eighth Circuit says that authority is divided among fifty states and the federal district courts around the country.

Worse, the Bell companies and GTE have found themselves a court ruling that could cripple—and certainly has delayed—the only viable avenue for the meaningful introduction of local competition to anyone but larger business customers. When a Bell company or GTE says that it has residential competition in its region, it is generally referring to competition made possible by total service resale—and precious little of it.

Total service resale is the only game in the market right now, and it is all too clear that such an approach will not work. Currently, many local companies cannot handle anything more than a trickle of orders for new competitors. The situation got so bad that AT&T has had to stop actively marketing local service in some states. AT&T could not risk its name on the kind of service the local companies were providing through total service resale. Even if the local companies could provide the operations support for total service resale, the margins for new competitors are unsustainably thin. Competitive offerings would be limited to "me too" copies of the incumbent companies' service. Thus, total service resale alone is not the quick solution that so many of us thought it would be.

To establish competition in local markets, the main focus is now on unbundled network elements, which enable fair competition and comparable wholesale costs in a market in which new entrants are still obliged to use an incumbent's facilities. The time travelers and the Eighth Circuit have placed the use of unbundled network elements on hold until we hear from the Supreme Court, because the Eighth Circuit ruled that the Telecommunications Act allows companies the opportunity to vandalize their own networks before they lease network capacity to competitors. The Eighth Circuit decided that *unbundled* meant ripped apart, even though in the history of telecommunications the word has never meant physically separated, but separately identified for reasons of pricing or cost.

Under the Eighth Circuit's interpretation, if AT&T leases such unbundled elements as a line to a home and some switching capacity, a new local customer could end up with a dead phone. The incumbent local company could rip the line from a house into the central office from a switch where it has been connected for years so that AT&T or another new local competitor would have to reconnect it—to the same place. That involves lots of unnecessary cost and intrusion into the incumbent company's central office just to reconnect connections that did not have to be severed in the first place. The Eighth Circuit decision appears to have left AT&T with a Works Progress Administration project without the social benefits, just the cost.

It is mind-boggling to say that an already assembled group of network elements could be pulled apart just to keep new competitors from entering the market. The Eighth Circuit's decision is all about protecting monopolies and standing in the way of customer choice. Fortunately, some states like Texas and Ohio are not waiting for the Supreme Court but have found separate bases for ordering an unbundled network element platform.

Connecticut and the Danger of Bigger Monopolies

The stakes are high and involve a potential reaggregation of monopoly powers. With the exception of cellular services, more than two-thirds of the telecommunications revenue generated in America still goes to the monopoly local exchange companies. From access charges from long-distance companies, intraLATA toll charges, and local service charges, the incumbent local exchange companies collect sixty-seven cents of every telecommunications dollar spent. Those long-standing monopolies want a shot at the remaining 33 percent before other telephone service providers can compete for their 67 percent. If those monopolies are the only firms that can offer a combined local and long-distance package, the specter of bigger monopolies is not an empty threat. A company that enters the long-distance market while it still has control over a local market could smother competition in both local and long-distance services.

Consider, for example, what has been going on in Connecticut. There, the local monopoly got into the long-distance market without opening its exchanges to competition. Doing so was easy for Southern New England Telephone. The firm could buy long-distance capacity from Sprint at a deep discount because the long-distance market is competitive. SNET then proposed charging a 48 percent premium to anybody who wanted to lease local residential capacity for resale in competition with the company. SNET offered a decent discount for business services but no ordering systems to get at it. While the potential local competitors went into arbitration, SNET entered the long-distance market with its local monopoly very much intact.

Evolving from having a monopoly in local service to having a monopoly grip on the ability to offer local and long-distance service together, SNET attracted customers from the original long-distance carriers. Indeed, the public has indicated over and over again that it wants one-stop shopping for local and long-distance services. As a result, SNET seized 40 percent of the long-distance market in Connecticut—not by offering especially deep discounts but by combining long-distance and local services with average industry discounts.

That experience did not push Connecticut to the head of the line as a state where AT&T would make long-term investments or deploy its most innovative technology. AT&T cannot do that where the local market is structurally unfair and allows no chance of building long-term value. AT&T makes such investments only where markets are open and the rules provide opportunities for delivering value to its customers and shareholders.

Now, BellSouth wants to make South Carolina into the next Connecticut. Doing so would be a great deal for BellSouth and a bad deal for the future of telecommunications in South Carolina, where consumers would be locked into an even stronger monopoly.

Recently, local exchange incumbents have been encouraging circulation of the fable that AT&T does not really want to get into South Carolina, Louisiana, and elsewhere—that AT&T just wants to exclude BellSouth from the long-distance market. BellSouth has claimed that it has opened local markets to competition but that nobody has entered. In fact, those markets are not nearly open. Even so, AT&T is trying to enter local markets in both South Carolina and Louisiana. In both states AT&T has valued local business customers on its digital offering on an outbound basis. AT&T wants to be able to offer all its customers a package that combines local, local toll, and long-distance services. BellSouth is reluctantly offering AT&T competitively trivial amounts of total service resale, not unbundled network elements, and operating systems inadequate to handle larger volumes. Those limited offerings do not add up to real competition.

Connecticut, South Carolina, and Louisiana offer prime examples of why AT&T has to hang tough on public policy. The United States has a comprehensive, procompetitive national telecommunications law. For the sake of America's telephone customers, AT&T has to see the Telecommunications Act through as it was written and ensure that it is enforced the way Congress intended.

Realizing that it is not natural for monopolies to give up power or to lease or interconnect their networks to competitors—even at a fair price—the framers of the act addressed the issue directly. They built into the act an incentive for at least 80 percent of the local exchange carriers to comply with the act's provisions for competitive local markets by offering the prize of entry into the long-distance market. Without that incentive, the competition AT&T led the public to expect will never happen or, at the very least, will be unconscionably delayed. Policymakers charged with implementing the long-distance entry provision of the act must be steadfast—and they have been. Their job will be difficult but not thankless, for the biggest prize is open markets and customer choices in places people had written off as "natural monopolies" just a few short years ago.

I remain optimistic about that outcome. The public wants choice more now than at any time in the history of telecommunications. Consumers have developed a taste for choice after a dozen years of vigorous competition in long-distance markets. They understand the relationship between competition and innovation, and they enjoy innovations like broadband services to homes, easy-to-use electronic commerce over the Internet, and video conferencing—not to mention better and more affordable basic services. That kind of innovation is the future—the near future if we make it happen.

Right now, the lack of competition in local service is like a glass door standing between the public and the future, and customers can see the innovations through that glass, even though the door is still locked. Full implementation of the Telecommunications Act is the key that will open that door.

This is no time to take a trip back to the future. The telecommunications industry should not undo what it already has. Instead, the industry should listen to the public and get competition out of the courtrooms and into the market where it belongs. Let the engineers, not lawyers, work the interfaces. That is within the industry's grasp, but time is wasting.

Delivered December 18, 1997.

11

Consumers Wanted Competition, but So Far It's No Contest

Richard D. McCormick

If "the era of big government is ending," it is certainly going out in style! The Center for the Study of American Business at Washington University in St. Louis reported that the federal government's spending on regulatory activity stated in constant 1992 dollars is at an all-time high of almost $15 billion. Fifteen billion just for federal regulatory agencies!

I was quite pleased to have the last word in this series on the Telecommunications Act of 1996. One of my associates offered the following overly simplified summary of some of the previous contributors' positions:

> Reed Hundt said, "The law is fine, 'cause the rules are mine! And I'm retiring just in time!" Ray Smith said, "Competition? We're here to try it. But if it's next door, we'd rather just buy it." Joel Klein said, "Competition should win. But if it should fail, we'll be here to put someone in jail." Bob Rowe said, "We need lots of data; no amount is too great. But we also need policy left with the state." And John Zeglis said, "Wouldn't it be nice to get out of court? We want to compete. But if they do? A tort!"

I received another summary from a group that has

not even spoken here—newspaper columnists and editorial writers. They all seem to think that what everyone in this industry is really saying is: "We just don't understand this horrible fight. When it hurts us, it's wrong. When it helps us, it's right!"

What I hope will be said about my message is: "McCormick spoke *not* for his firm alone; he urged that we think of the person with a phone!"

Certainly, amid all the furor over the Telecommunications Act and its implementation, the interest group from which one has heard the least is customers—telephone customers, cable TV customers, Internet users—the people who know that those wires and airwaves are lifelines with which people increasingly earn, learn, manage their lives, enrich their free time, and even get health care.

Customers are the real interest group in telecommunications policymaking. The question we all need to answer—whether we are in government, the telecommunications industry, broadcasting, or software—is, Are we evaluating everything we do in terms of meeting the customer's desire for more services? For more providers? For a future in which wires replace tires as the primary mode of getting things done? Or are we—all of us—excessively preoccupied with "rules" and "turf"?

Clearly, the intent of the Telecommunications Act of 1996 was to open all the turf to all the players. The act provided a foundation for growth in a sector that is already one-sixth of the U.S. economy. The act provided a framework on which companies could add the "finishing touches" to suit Americans' individual needs and deliver those solutions to customers' doorsteps.

As I looked at the act twenty-three months ago, I did not envision the overzealous building inspectors. Congress intended that Americans quickly get more choices of providers, as well as new and useful services, including high-speed data and two-way video services—at home, as well as at work. Congress also intended that all customers benefit from price competition.

Congress and most of us in the industry were optimistic, and we should remain so. While some of the erroneous interpretations of the law have required clarification in the courts, that is not where customers want the industry to fight the battle. While U S WEST has been a party to some of those court challenges, everyone in the telecommunications industry needs to spend less time trying to beat the competition in the courtroom or hearing room and more time trying to be the competition in the laboratory, switching center, and sales room. The more valuable services the industry develops, the more customers will see the benefits of an open, unregulated marketplace.

What is open so far? Wireless, video, and Internet access. But certainly not long-distance markets. Long-distance service remains the elusive "no-Bell prize." What about local telephone services? Are they open to competition? A firm wanting to enter the phone business in Denver does not need to own a single wire. The firm can send a service order from its computer to U S WEST's computer, get confirmation within hours, and get the service as fast as U S WEST would do it for itself. The entering firm does not have to prove first that *its* business has competition. The firm certainly does not have to open its facilities to its competitors! It does not have to pass a federal agency's cumbersome "checklist." Nor does it have to hire a corps of lawyers. U S WEST's network is definitely "open for businesses"—for entrants to use as if it were their own.

In the two years since the Telecommunications Act was passed, U S WEST has invested more than $400 million in network "interfaces," which are the hardware and software to allow other local exchange companies to connect with, resell, and compete with our services. U S WEST has assigned more than 600 employees to working with local competitors to facilitate their use of its network, in the name of "competition." U S WEST has in place 244 state-approved interconnection agreements with some eighty-seven telecommunications companies in its fourteen-state territory. U S WEST has another 143 agreements pending and is negotiating 198 others. Currently, U S WEST is processing more than 4,000 service orders a month for local competitors.

U S WEST is not alone. Nationally, the Bells and GTE have spent more than $4 billion to open their networks to new entrants in the local telephone market. Unfortunately, those new entrants seem to be interested mostly in business customers—not residential. Far too often, the information highway seems to end downtown and in the larger suburban business parks. But it never seems to "go home." Should regulators not make note of that?

U S WEST has made substantial expenditures to accommodate local competitors. Yet some of them call U S WEST's gateways "roadblocks" and claim that the company has not opened its doors to competition. Could it be that those competitors want to find roadblocks so that they can avoid all the expense of providing local residential service while keeping all the profits from residential long-distance service?

Although many have referred to the "battles of the giants" over the telecommunications pie, few have noted the progress the industry has made in *enlarging* the pie. Dozens of companies are making competition work—at a profit to them and to their customers. Every day, U S WEST works with small companies that are finding ways to use its facilities with their products and services and their greater regulatory freedom to add value for customers.

For example, one Midwestern company is buying lines from U S WEST under a multiline business tariff. That new vendor is reselling those lines with more liberal credit terms than U S WEST chose to extend. The vendor is providing a customized bill and more customer instruction and handholding. The results are more services and a bigger market than previously existed.

U S WEST has another "resale" customer that concluded that certain kinds of sales offices are willing to pay a premium for a system that gives each salesperson an ID code for long-distance calls. That allows the office to know which salesperson is responsible for which costs. The vendor buys phone service from U S WEST and adds the ID feature. The only loser is the salesperson who used to beat his way out of paying for some of his calls. What U S WEST is experiencing is open competition that works—new services, new vendors, new economic growth.

Meanwhile, U S WEST is working hard to be the competition. U S WEST has extended caller ID and voicemail to thousands of new customers in the past few years and has introduced a first-in-the-nation "one-number" service, so that when customers are not at home, their calls can follow them to their wireless PCS phones.

In Phoenix U S WEST has wired a new subdivision with fiber so that residents can get phone calls, TV, and high-speed Internet from a single network. Elsewhere, U S WEST has upgraded to digital subscriber lines to bring the future to more homes. U S WEST foresees more and more exciting services for all of its customers.

The trouble is that regulations that focus excessively on the past hamstring U S WEST. The FCC and the Justice Department have mistakenly viewed today's market as "static" and have made a lot of rules with a view to "dividing up the pie"—not to mention overseeing the slicing. What Congress had in mind in 1996 was to provide the telecommunications industry with the environment and the incentives to bake a bigger pie. Congress wanted to give the American people and the American economy the tremendous advantages of a telecommunications system that is limited only by scientists' and entrepreneurs' imaginations. And, of course, Congress wanted to see those benefits at home, as well as in the workplace. That is what many firms in the industry wanted. Most important, that is what customers wanted.

But that is *not* what the industry got. The rules the local telephone operating companies have been given and the "competition" those firms have seen create a major question: Where is the incentive for a regional Bell operating company or GTE to build a combined network to deliver voice, video, and high-speed data to every home? What such firms have, in fact, is a tremendous disincentive because the company that tries to build that network does so at the frightening risk of watching some other firm earn the return on that huge investment.

Where is the opportunity to offer both local and long-distance services? The long-distance companies have that opportunity, but they do not seem very eager to act on it. But, unless the Texas ruling stands, regional Bell operating companies cannot sell interstate long-distance service within their operating regions. Nor can regional Bell operating companies bundle that service with other services—even though 70 percent of customers tell the Yankee Group that they want one-stop shopping.

Every day, customers ask U S WEST why the firm is not able to provide that one-stop service. U S WEST is "able" to do so but is not allowed to do so. So telephone customers go through a lot of extra hassle dealing with multiple vendors just to accommodate the federal rules. Is that what Congress had in mind for American families and businesses?

Is the Telecommunications Act a good thing? It certainly has been good for customers of wireless communications. They got more services, better quality, lower prices, and the likelihood of more to come. The Telecommunications Act has been at least somewhat good for long-distance customers as more companies, including the Bells out-of-region, continue to increase choices and reduce prices. And, of course, the act has been good for the long-distance companies, which have continued to keep local exchange carriers out of their markets.

The law has stimulated increased choices for cable service. Consumers will see more as new technologies come on-line. The law has been good for people in Southern New England Telephone Company's territory. When that local exchange company was allowed to enter the long-distance market, suddenly everybody was competing for everything, and every customer benefited. Finally, the new law made it clear that nobody is playing "Monopoly" anymore—despite what Americans may have heard from those who like to pretend that the Bells are still in that old game.

The law itself is still generally "good." What is *not* good is the way the rulemakers have implemented it. Who rules? The federal government or the states? The law made the answer clear, but the battling bureaucracies have muddied it up.

What does *interconnection* really mean? The law made that clear, but the FCC made a sharp left turn. The Supreme Court may weigh in on the matter.

What does *universal service* really mean? The law said "a phone in every home" with a "cost-sharing pool" to keep the most remote homes connected. But some companies object, because all they want are the easy profits, downtown, while they leave some other firm entirely responsible for their own mother's telephone!

What does *opening local markets* really mean? To me, $4 billion worth of "opening" looks pretty real. Virtually any firm can sell local phone services on U S WEST's network today. Why is U S WEST unable to sell long-distance service on the competitors' networks?

The implementers of the law are singling out U S WEST for unfair restrictions, and U S WEST resorted to a court challenge of the law itself when confronted with that discrimination. I do not have a law degree, but I do have a neighbor who would *like* one-stop shopping for local, long-distance, and wireless phone services. He would like a choice of where to buy those services. He would like more TV channels, faster Internet access, more time-saving services, and price competition for "all of the above." He and I would both like to see real competition at every customer's doorstep.

The Telecommunications Act was passed in good faith. Thousands of Americans have worked hard to implement it fairly. Now is the time to finish the job. It is time to make the rules match Congress's intent and consumers' desires. It is time to reduce that $15 billion Americans spend on regulation. It is time to give customers the full benefit of the Telecommunications Act—voice, video, and high-speed data services.

We cannot regulate our way to competition. We cannot litigate our way to competition. But given the opportunity, we can *innovate* our way to competition. And we must—for our customers and our country. It is time to bring the Information Age "home"—where it belongs.

Delivered January 20, 1998.

The American Enterprise Institute for Public Policy Research

Founded in 1943, AEI is a nonpartisan, nonprofit, research and educational organization based in Washington, D. C. The Institute sponsors research, conducts seminars and conferences, and publishes books and periodicals.

AEI's research is carried out under three major programs: Economic Policy Studies; Foreign Policy and Defense Studies; and Social and Political Studies. The resident scholars and fellows listed in these pages are part of a network that also includes ninety adjunct scholars at leading universities throughout the United States and in several foreign countries.

The views expressed in AEI publications are those of the authors and do not necessarily reflect the views of the staff, advisory panels, officers, or trustees.

Research Staff

www.ingramcontent.com/pod-product-compliance
Lightning Source LLC
Jackson TN
JSHW011940131224
75386JS00041B/1470
9780844740942